# Sustainable Aviation

**Series Editors**

T. Hikmet Karakoc, Faculty of Aeronautics and Astronautics
Eskisehir Technical University
Eskisehir, Turkey

C Ozgur Colpan, Department of Mechanical Engineering
Dokuz Eylül University
Buca, Izmir, Turkey

Alper Dalkiran, School of Aviation
Suleyman Demirel University
Isparta, Turkey

The Sustainable Aviation book series focuses on sustainability in aviation, considering all aspects of the field. The books are developed in partnership with the International Sustainable Aviation Research Society (SARES) and include contributed volumes consisting of select contributions to SARES international symposiums and conferences, monographs, and professional books focused on all aspects of sustainable aviation. The series aims at publishing state-of-the-art research and development in areas including, but not limited to:

- Green and renewable energy resources and aviation technologies
- Aircraft engine, control systems, production, storage, efficiency, and planning
- Exploring the potential of integrating renewables within airports
- Sustainable infrastructure development under a changing climate
- Training and awareness facilities with aviation sector and social levels
- Teaching and professional development in renewable energy technologies and sustainability

Can Ozgur Colpan • Ankica Kovač
Editors

# Fuel Cell and Hydrogen Technologies in Aviation

*Editors*
Can Ozgur Colpan
Department of Mechanical Engineering
Dokuz Eylül University
Izmir, Turkey

Ankica Kovač
University of Zagreb
Faculty of Mechanical Engineering
and Naval Architecture
Zagreb, Croatia

ISSN 2730-7778 ISSN 2730-7786 (electronic)
Sustainable Aviation
ISBN 978-3-030-99020-6 ISBN 978-3-030-99018-3 (eBook)
https://doi.org/10.1007/978-3-030-99018-3



This Springer imprint is published by the registered company Springer Nature Switzerland AG
The registered company address is: Gewerbestrasse 11, 6330 Cham, Switzerland

# Preface

The aviation sector is significantly responsible for greenhouse gas and air pollutant emissions. Despite the efficiency improvements in aircraft engines, the industry growth rate is higher; thus, the emissions continue to increase annually. Many countries and organizations have already agreed on reducing these emissions substantially in the near future. Hence, using clean fuel and technologies in aviation plays a key role in achieving these goals. Electric aircraft using batteries, provided that electricity used to charge the batteries is produced from renewables, biofuel or synthetic fuel-fed engine-powered aircraft, and hydrogen-fed fuel cell-powered aircraft seem to be the most promising options in this regard. Significantly, the use of hydrogen energy is increasing rapidly in the aviation sector and several new prototypes and products are released continuously.

This book gives an overview of the current status and future directions on the use of hydrogen and fuel cell technologies in aviation and their potential to reduce the environmental impact of aircraft. Their technological and economic feasibility and requirements in the aircraft design and airport infrastructure are discussed. Different hydrogen storage technologies that can be used in aircraft, including compressed gaseous hydrogen, cryogenic compressed hydrogen, cryogenic liquid hydrogen, metal hydride hydrogen, chemical hydrogen, organic hydrogen, and adsorption hydrogen storage, are covered. As liquid hydrogen is quite promising, especially for long-range flights, a separate section is devoted to this fuel to show the status and trend of its use in aviation. Infrastructure, logistics, and safety requirements to operate hydrogen-powered aircraft at an airport before they perform commercial flights are also handled in this book. Applications of fuel cells in unmanned aerial vehicles and passenger aircraft and several topologies that can be used, are addressed. Optimizing the sizing and energy management of fuel cell-based aircraft powertrains is also important in terms of achieving long range and operating with low cost. This book also describes a method to obtain an optimized energy management strategy for fuel cell-powered aircraft. The use of solid oxide fuel cells as auxiliary power units and a hybrid propulsion unit in conjunction with gas turbines is also discussed in detail.

This book is expected to guide researchers, scientists, engineers, and graduate students in the selection, design, and assessment of hydrogen storage and fuel cell technologies for various aircraft types. We would like to thank the Sustainable Aviation Research Society (SARES) for giving us the opportunity to publish this book, Springer editorial team for helping and guiding us in each step of the preparation of the book, and all chapter authors and reviewers for their excellent contributions in the success of this book.

Izmir, Turkey
Zagreb, Croatia

Can Ozgur Colpan
Ankica Kovač

# Contents

# Hydrogen Storage Technology for Aerial Vehicles

Dirk Kastell

## 1 Introduction on Hydrogen Storage

The use of hydrogen in aerospace has a long history in the last century, especially when looking at space applications. Since the early 1930s, hydrogen was extensively used, as it is the only fuel capable to power rockets into earth's orbit. Therefore, today, the use of hydrogen as propulsion fuel is in space application still the standard. Nevertheless, for the civil aviation industry, the use of hydrogen was always experimental and is still under investigation. Even when the requirements for space applications are high, they are different of those in aviation. An aircraft can be in service for over 30 years and will transport in this time thousands of passengers, compared to the one-time use of a space rocket and the limited transport capabilities. Therefore, hydrogen storages for the aviation industry have to be qualified to the more stringent aviation certification requirements, which include intensive vibration and thermal fatigue testing for the lifetime period. Today, besides the classical way of gas storage, there are other ways to bunker hydrogen. Especially, in the last two centuries, many applications have been investigated. The growing popularity to use fuel cells for mobility also pushed this technical interest on these developments. This chapter on hydrogen storage technologies in aviation is intended to give an overview of the available methods.

D. Kastell (✉)
Hamburg, Germany
e-mail: dirk.kastell@gmx.de



C. O. Colpan, A. Kovač (eds.), *Fuel Cell and Hydrogen Technologies in Aviation*, Sustainable Aviation, https://doi.org/10.1007/978-3-030-99018-3_1

## 2 History of Hydrogen Storage

Looking at the limited history of hydrogen aircraft, there are the MARTIN B-57B in the 1950s and the TUPOLEV TU-155 in the 1970s, setting the standard in their century. Both were using liquid hydrogen and both were not commercialized due to technical barriers, but mainly because of the affordable price. Focusing only on hydrogen storage for flying vehicles, there are older examples, which are the Zeppelins. They were not flying with hydrogen (they used diesel), but they used hydrogen as light gas to fly [1]. At this time, helium was expensive and had limited availability. For example the famous LZ129 Hindenburg Zeppelin stored 190,000 m$^3$ of hydrogen in its gas cells (Fig. 1). At that time, the cells were made of gold bat skin, but on that Zeppelin two plies with an in-between layer of a gelatinous substance were used. For the gelatine, various types of natural and artificial rubber, as well as various other, mostly synthetic substances such as oppanol

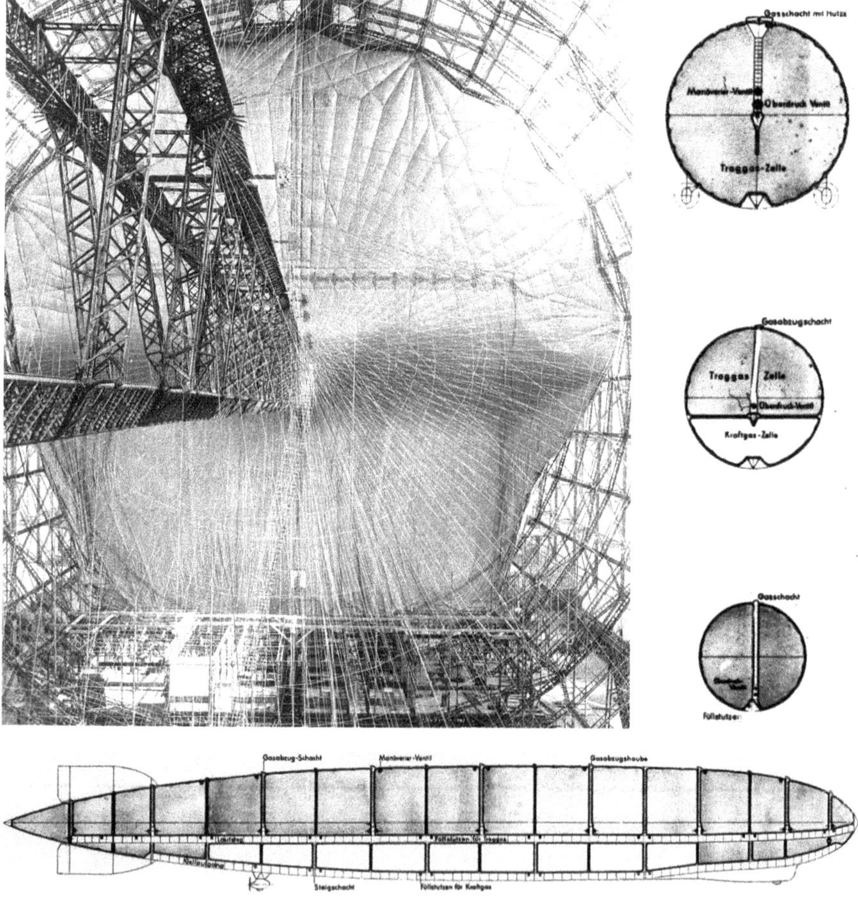


**Fig. 1** Hydrogen gas tank of the Zeppelin Hindenburg LZ129 (*UL* installation, *UR* cut views at three positions of the aircraft shown in cut view below)

(polyisobutylene) were used as sealants. The gas permeability with this technology was only 1/m$^2$ in 24 h, which is still challenging today, having in mind the free surface of 56,878 m$^2$ and a cell weight of 203 g/m$^3$. At that time, the technology requirements were the same as today; the hydrogen gas tanks had to be lightweight and tide. Especially because of the permeability of hydrogen as the lightest element and a gas of diatomic molecules with the formula $H_2$, special care and technologies have to be used to keep the gas in its storage. A heavy metal storage tank is the easiest solution, but in commercial aviation, it leads to a decrease of the payload capacity. Therefore, even today, much research is investigated into the development of lightweight materials like carbon fibre or thermoplastic. Looking at the typical storage condition at high pressure for gaseous hydrogen or low temperature for liquid hydrogen, this creates additional technical challenges.

Whereas in the past the use of hydrogen was mainly related to the application of propulsion or with the intention to have a fast-flying aircraft like a supersonic aircraft, today the use of hydrogen is driven by the idea to decarbonize the aviation industry. So besides the intention to burn hydrogen in engines, which will lead to NOx emission, hydrogen can be used with a fuel cell as an alternative electric energy generation source to fly an aircraft with no emission at all. If hydrogen is also produced without carbon emission, e.g. generated by sustainable energy (solar, wind or bio), the propulsion technology is emission free.

## 3 Hydrogen Storage Technologies

There are many ways to store hydrogen for later use. From the thermodynamic point of view, it has to be remembered that hydrogen boils at −252.77 °C (20.33 K). The melting point is at 259.1 °C, and the critical point is at −240 °C at 1 atm. The density is 0.0899 g/l at standard atmosphere. Liquid hydrogen has a specific gravity of 70.99 g/l, which is 790 times the one of gaseous hydrogen. The volume-related energy density of liquid hydrogen is about 1/4 that of gasoline and about 1/3 that of natural gas. The weight content of hydrogen in water is 11.2%.

The typical hydrogen is gaseous at standard atmospheric condition, and therefore, it is most times stored in this form at different pressure conditions. Today pressures of 350 bar and 700 bar have been established for different usages. Especially in the mobility sector, it is often used in this form as it is comparable to other gas storage systems used in the automotive industry. Nevertheless, there are various other ways to store hydrogen, e.g. as cryogenic compressed or in liquid condition. Other ways of hydrogen storage are using different kinds of media like adsorption material and metal hybrid. Nevertheless, most storages use a vessel. Another way to store hydrogen is to convert it to a different chemical compound like methane or ammoniac. These compounds can then be stored and transported very easy, and hydrogen can be extracted for its final use at the place it is needed. Nevertheless, this kind of change is not always possible at every place. The uses of the last kind of storage conditions are usable for many industries. However, having aviation in mind, because of the additional safety risks going along with the regeneration of hydrogen, today only the physical storage of hydrogen in a vessel is in the focus of the development in the aviation industry.

Specifying different types of hydrogen storage, there are six types of conservation. All storage methods have their advantages and disadvantages, each of which qualifies them for different tasks. In practice, there are also mixtures of different types used:

- Compressed gaseous hydrogen
- Cryogenic compressed hydrogen
- Cryogenic liquid hydrogen
- Metal hybrid hydrogen storage
- Chemical hydrogen storage
- Organic hydrogen storage
- Adsorption hydrogen storage

Table 1 gives an overview and gradation about the different storage technologies in temperature, pressure and density. These values are typical figures and only shall enable a comparison between the methods. Also the list is not completed as new methods are developed, and many investigations are going on to push the hydrogen density especially for the new technologies to higher numbers.

## *3.1 Physical Storage Technology*

The physical storage of hydrogen is the basic technology of this medium in the industry. It is of normal use today, as it is easy and of low cost compared to other technologies, which are in some cases still under investigation and expensive in use. The physical way of storing hydrogen does not need any kind of forward or backward conversion, which requires additional systems and energy. Figure 2 shows a thermodynamic diagram of hydrogen density over temperature in relation to pressure. Especially marked are the four main physical storage conditions which are used today for hydrogen in a tank. Typically compressed hydrogen $CH_2$ is used at 700 and 350 bar. It has to be noted that the stored density rises with increasing pressure and decreasing temperature. The highest density changes occur near the critical point (CP). Therefore, the use of liquid hydrogen ($LH_2$) and cryogenic compressed hydrogen ($CryoCH_2$) are of high interest for the aviation industry.

### Compressed Hydrogen

Compressed storage is always referred when a type of gas is stored under higher pressure than normal pressure. The pressure for the storage depends mainly on the area of application and the pressure level needed. In the case of stationary use, like underground cavern storage facilities, hydrogen is stored at a maximum of up to

**Table 1** Differentiation of storage conditions for hydrogen [2, 6, 15]

| Type to store | Form of storage | Temperature | Pressure | Medium | $H_2$ density (kg/m$^3$) | Storage density (mass%) | Comments |
|---|---|---|---|---|---|---|---|
| Physical storage technology | Pressurized | 293 K/20 °C | 350 or 700 bar | Gaseous | 25–42 | ~4 | Low quantities, serial standard |
| | Cryo compressed | 50 K/−223 °C | 350 bar | Gaseous | ~80 | ~5 | High storage volume, heavy and expensive |
| | Cryo liquid | 37 K/−236 °C | 3–5 bar | Liquid | ~65 | ~10 | High quantities, low pressure, standby, boil off |
| Hydrides storage technology | Metal hybrid | Ambient | Ambient | Metal alloy powder | ~45–150 | ~2 | Heavy, expensive, complex, safe |
| | Chemical | >373 K/>100 °C | Ambient | Solid or powder | ~ 49–105 | 1–18 | Complex |
| | Organic | Ambient | Ambient | Liquid | ~1 | ~6 | Safe handling heat to release |
| Adsorption storage technology | Carbon | <77 K (best)/<−196 °C | >100 bar (best) | Solid | ~18 | 1–2 | Storing needs high energy |
| | MOFS | <253 K/<0 °C (best) | >100 bar (best) | Solid or powder | ~5–45? | 1–5 | Experimental expensive |
| | Zeolite | <77 K (best)/< −196 °C | >100 bar (best) | Solid or powder | ~11–122 | 1–2 | Safe storage heavy |

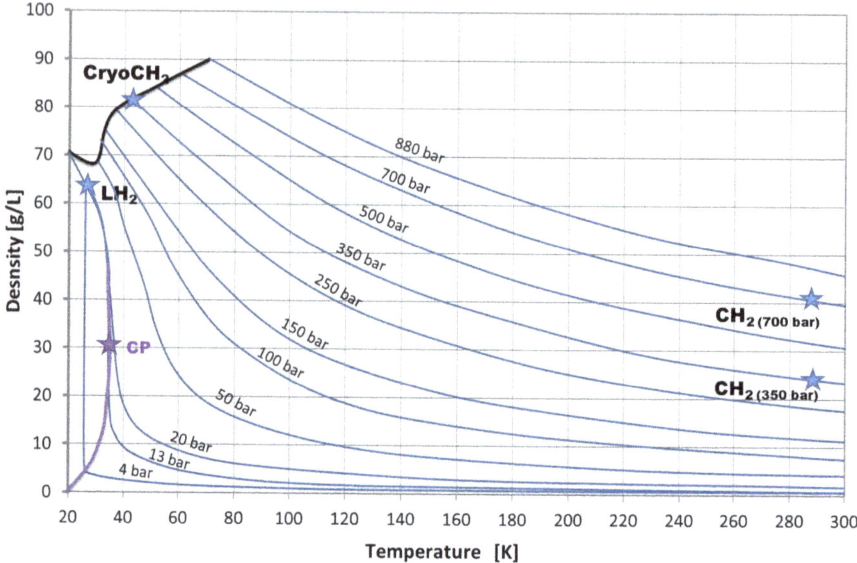


**Fig. 2** Diagram of storage conditions of pure hydrogen [2, 3]

50 bar, as the volume usually plays less of a decisive role. Low pressure also means less loss. In mobile applications, hydrogen volume is essential, and accordingly, higher pressures are used for storage. Higher pressure means more hydrogen and significant distance to drive. However, at higher pressure, more safety precautions have to be taken into account. Accordingly, there are today two different pressures used in the mobility sector: 350 and 700 bar. Most automotive manufacturers prefer pressure tanks at 700 bar. At this pressure level, fuel cell vehicles can achieve ranges comparable to gasoline vehicles. As the time needed to refuel is comparable with gas fuelling, it is felled like no change. For other vehicles like forklifts, which can easily be recharged at any time when needed, 350 bar is used. This pressure level is also available for bottled hydrogen, and so it is commercially available.

## Cryogenic Liquid Hydrogen

Liquefied storage of hydrogen has the highest storage density in relation to the pure storage volume. The intention behind this use is to minimize the storage size of the medium. The advantage of liquid storage is not only the size, but also the storage at low pressure of around 3 bar. The disadvantage is that hydrogen becomes only liquid at −253 °C. Therefore, the storage systems for liquid hydrogen have to be very specialized. These cryogenic tanks need reliable insulation to keep the vaporizing due to heating low. Lately, a new storage at a slightly higher pressure and lower temperature than ambient of liquid hydrogen is used, which is called sLH2—subcooled liquid hydrogen [4]. The intention is that the subcooled hydrogen can help as a buffer to liquefy boiled off hydrogen again after refuelling.

Compared to the standard kerosene used in aviation, hydrogen has only a third of the energy of this medium. Today, the use of liquid hydrogen is anticipated by the aviation industry to overcome the usage of large vessels in the aircraft which badly influence the aerodynamic or the payload of the aircraft. That is why liquid hydrogen is also used for centuries as rocket fuel in space travel. Liquid cryogenic hydrogen is also used for storage on ground for delivery by truck for reasons of space. It can then be evaporated into gaseous hydrogen at the filling station. In practice, liquid hydrogen usage has disadvantages compared to the pressure variant. For example, evaporation losses cannot be completely avoided during longer service lives of vehicles. Moreover, the energy required to provide liquid hydrogen is higher than for pressurized hydrogen. Today also, the industrial capacity to produce liquid hydrogen is not high and will be one of the challenges in the future for the usage of this medium.

#### Cryogenic Compressed Hydrogen Style

Cryogenic compressed hydrogen storage refers to the storage of hydrogen at cryogenic temperatures in a pressurized vessel (nominally at 350 bar), in contrast to current cryogenic vessels that store liquid hydrogen at near-ambient pressures. Cryogenic compressed hydrogen storage can include liquid hydrogen, cold compressed hydrogen, or hydrogen in a two-phase region (saturated liquid and vapour). With cryogenic compressed hydrogen even higher densities than with liquid hydrogen can be reached [3]. But this condition goes along with the design requirements not only to have a good insulation of the vessel but also to apply the precaution for a high-pressure accumulator, which means in most cases thicker walls and higher weight. So where on the one side, the cryo-compressed storage system has the potential to meet the needed volumetric capacity in the automotive industry, it has a high-volume manufacturing cost [5]. But especially the weight is a drawback for the aviation industry and is therefore not considered for usage [18].

### 3.2 *Hydrides Storage Technology*

One other different way of storing hydrogen is the use of chemical hydrides, which means hydrogen is stored as a combination of another chemical element. This binding is not fixed and can be released at any time, when a special treatment, e.g. heat or pressure, is applied. In some cases, even this is not required. The overview below gives the formula of most hydrogen chemical compounds with their state at standard atmosphere. Many of the chemical hydrides used today in research are using those combinations as a baseline. This application is a wide open field of experimental examinations, which are still going on today. Prior to the 2003 timeframe, most material-based hydrogen storage technology development had focused on reversible interstitial metal hydrides. Nevertheless, some elements have already left the experimental stage. In Fig. 3, the stored hydrogen mass for different elements is

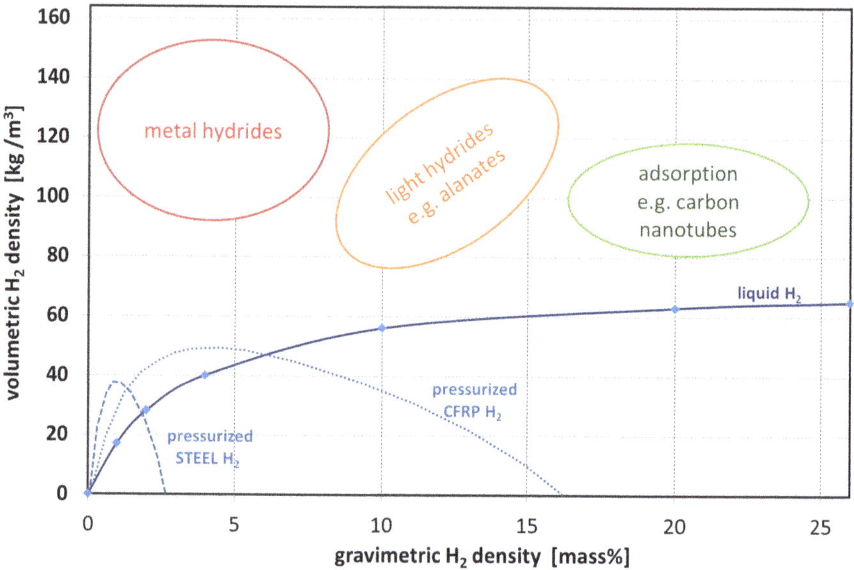


**Fig. 3** Stored hydrogen per mass and per volume in comparison of physical hydrogen storages with metal hydrides, carbon nanotubes, and other light hydrides [6]

compared to the classical pressurized storage in steel or carbon fibre (CFRP) housing or as a liquid. Some elements can take more hydrogen than their own physical density [11].

## Metal Hybrid Hydrogen Storage

In this storage technology, certain metal alloys with standard metals like titanium and iron, lithium hydride ($LH_2$), or magnesium and sodium alanates are used. They have the capacity to absorb large amounts of hydrogen [7]. In practice, hydrogen storage containers are filled with these metal powders and absorb and release hydrogen like a sponge. For example when hydrogen is bound by sodium aluminate, the hydride complex sodium aluminium hydride is formed. Other well-known elements are Solid Ammonia Borane, Metal-Boron-Nitrogen and Hydrolysis of sodium borohydride ($NaBH_4$). All of those elements have the capability to store hydrogen, but as well also disadvantages. When a metal hydride is 'filled' with hydrogen, it emits heat. To retrieve the hydrogen back from the storage, the hydride has to be heated again. In terms of volume, very good values of storage capacity of these systems are obtained. Unfortunately, the metal memories are comparatively heavy, so that their use for mobile applications is limited. In addition, these storage systems are comparatively expensive due to the high material costs. Metal hydride accumulators have clear advantages in handling and safety. The hydrogen is only released by heat, which means that the hydrogen remains bound in the event of damage to the storage tank. They work almost at normal pressure, show no evaporation losses and also have a cleansing effect on the hydrogen. Therefore, today these memories are used mainly commercially in submarines. Lately, a different usage has been found, where the magnesium hydride ($MgH_2$) is not stored in a vessel but is generated and stored as grease. This has the advantage that the hydrogen can be handled easily and is still contained as it is only released at 300 °C, which occur for example during combustion [8].

## Chemical Storage of Hydrogen

As shown in the Table 1, there are many chemical compounds that react with hydrogen and serve therefore as chemical storage. Chemisorption allows the formation of stable compounds which can store hydrogen at ambient conditions, so they are interesting candidates for mobile applications. But chemisorption suffers sometimes from reversibility, because not all compounds can regenerate the pure hydrogen. The behaviour of these compounds for storing hydrogen has been intensively investigated [9].

Ammonia is the most favoured compound today; it is a compound of nitrogen and hydrogen with the formula $NH_3$. It is a stable binary and the simplest hydride [10]. Ammonia is a colourless gas with a distinct characteristic of a pungent smell. It is also a building block for the synthesis of many pharmaceutical products and is used in many commercial cleaning products. $NH_3$ boils at −33.34 °C for ambient pressure, so the liquid must be stored under higher pressure or at low temperature. Ammonia has a high capacity for hydrogen storage of 17.6% mass, based on its molecular structure. However, in order to release hydrogen from ammonia, significant energy input as well as reactor mass and volume are required. Other considerations include safety and toxicity issues. Although common in nature and in wide use, ammonia is both caustic and hazardous in its concentrated form. Another disadvantage is the incompatibility of polymer electrolyte membrane (PEM) fuel cells in the presence of even trace levels of ammonia (>0.1 ppm). Nowadays, first, experiments exist for fuel cells driven by ammonia.

Another well-known element to bind hydrogen is methane, which is often referred to as a substitute for a hydrogen media. Methane is a combination of hydrogen with carbon to $CH_4$. It boils at −161.5 °C, so it is gaseous at standard atmosphere as well. In most cases, methane is burned directly, because to release the hydrogen from methane high temperatures (900 °C for steam reforming and 1100–1300 °C for thermal release) are needed depending on the process used.

## Organic Hydrogen Storage Style

One other interesting hydride method of storing hydrogen is the use of synthetic aromatic-based heat transfer oils that are generally used in bakery systems and other high temperature applications. Here the heat transfer oil acts as a liquid organic hydrogen carrier (LOHC). LOHCs are organic compounds that can absorb and release hydrogen through chemical reactions. LOHCs can therefore be used as storage media for hydrogen. The hydrogen is stored inside the liquid hydrogen carriers via a catalytic reaction. The liquid now has a low viscosity and looks like water. After the hydrogenation by the temperature of around 300 °C, the viscosity increases again and the liquid looks like honey.

Since the (optimal) LOHC is liquid at ambient conditions and shows similar properties as crude oil based liquids (e.g. diesel, gasoline), it can easily be handled, transported and stored. When loaded with hydrogen, LOHC is flame-retardant, which makes it a safe transport medium for hydrogen to the location of use where the hydrogen can be unloaded from this carrier liquid. Today, the most promising LOHC candidates are dibenzyltoluene for energy-transport and energy-storage as well as N-Ethyl-Carbazole for mobility applications [11].

## *3.3 Adsorption Storage Technology*

### Carbon Hydrogen Storage

One other way of storing hydrogen is a similar use like the metallic hybrid solution. In this case, materials with a large inner surface area and suitable pore size, which prefers to attach certain gases, are used. Porous carbon is a long-known material with those suitable properties [12]. This carbon is a synthetic carbon modification containing very small graphite crystallites and amorphous carbon. The operational dynamics can be over a wide pressure and temperature range, but it requires complex storage management. There is a thermal method with a treatment at 700–1000 °C in the presence of oxidizing gases such as $CO_2$ steam or air and the chemical method with a treatment at 500–800 °C in the presence of dehydrating chemical substance. All of these treatments are only rudimentary developed today. Typically the carbon is produced in nanotubes or nanofibers, which are used to fill up a storage vessel. The reversible hydrogen sorption process is based on physisorption, and the amount of adsorbed hydrogen is proportional to the surface area of the nanostructured carbon. Typically a storage of lower than 1% mass can be achieved, but when the storage condition is changed to low temperature 77 K and high pressure >100 bar higher storage density of at least 2% mass is achieved [13]. Nevertheless, this storage condition is again leading to higher efforts for the storage vessel similar to the one for cryogenic compressed hydrogen.

### Hydrogen Storage in Zeolite

Zeolites are also well-known storage materials. They can be of natural and synthetic origin and are crystalline aluminosilicate compounds with a cavity- and channel-type pore structures, used to store the hydrogen. Their pore structure offers a high internal surface area, nevertheless at ambient condition only a hydrogen uptake of <1% mass is reached [13]. To improve the storage mass of hydrogen like for carbon, the pressure and temperature have to be raised to higher values. At the same time, also different substitutions like lithium borohydride $LiBH_4$, used also as metal hydride, can be used to improve the storing capacity [14].

Nevertheless zeolites will probably never be considered as an ideal storage medium for hydrogen in mobile application, as they are all relatively heavy compared to carbon or MOFs (where carbon, nitrogen and oxygen make the bulk of the structure). But they have potential as a cheap bulk stationary storage, because of their thermal and chemical stability to store hydrogen safely.

### MOFs as Hydrogen Storage

The last named storage material is the so-called metal organic frameworks (MOFs). They are a new class of materials, named coordination polymers, which are synthesized to allow an influence on the internal surface and porosity. Coordination polymers reach internal surfaces of well over 3000 $m^2/g$. Therefore, a potential for gas storage, in particular carbon dioxide, methane or hydrogen, is attributed to them and partly observed. At ambient conditions, storage of 2.5% mass could be achieved. Therefore, the chemical tuning of the MOFs is investigated to improve the binding capacity of hydrogen [15]. Also for that material, the use of cryogenic temperatures and pressures of up to 90 bar can lead to a reversible hydrogen storage capacity of around 10% mass. To date, however, there is no functioning storage system made of these materials that would have storage properties better than active porous carbon.

## 4 Design of Hydrogen Storages

The design of hydrogen storages always depends on the storage technology chosen. Nevertheless, in most cases, a metal case is used, because when storing pure hydrogen, the containment of the medium in respect of the permeability of the medium has always to be taken into account. In the case of chemical or adsorption storage the used components have to be contained in a vessel, whereas some of them like ammonia are even hazardous. Because of the experimental status with almost all alternative storage technologies and the risks going along with this, the use of physical hydrogen storage systems are the main focus of the commercial aviation industry. It has to be noted, that in some special cases like unmanned flight vehicles (e.g. drones) the advantages could be higher than the risk to go over to alternative technologies. Nevertheless, the following chapters will only address the physical storage technologies for commercial aviation.

When designing hydrogen storages, the following points have to be taken into account:

- System level architecture
- Storage control, monitoring and indication
- Hydrogen storage design and installation

- Tank venting and operation management
- Redundancy and failure mode reconfiguration

## 4.1 System-Level Architecture

Pressure vessels are organized in five different types which are related to the material and design used:

- Type I: All-metal construction, generally steel.
- Type II: Mainly metal with some fibre overwrap in the hoop direction, mostly steel or aluminium with a glass fibre composite; the metal vessel and composite materials share equal structural loading.
- Type III: Metal liner with a full composite overwrap, generally aluminium, with a carbon fibre composite; the composite materials carry the structural loads, the liner is for the tightness.
- Type IV: An all-composite construction, polymer (typically high-density polyethylene) liner with carbon fibre or hybrid carbon/glass fibre composite.
- Type V: Liner less, all-composite construction.

Hydrogen tanks for storing liquid hydrogen in aerospace are in many cases made of metal. For space applications, this solution is still in use, because of non-reuse and costs. But for commercial aviation, the focus is on the design of a $LH_2$ storage system made of light carbon fibre-reinforced materials. Using these components, a significantly improved gravimetric index for the whole storage system can be achieved. The drawback of using these materials is that the laminate quality or laminate architecture is of particular importance for the permeability. Different test specimens, different semi-finished products made of glass, carbon, fibres with polymer or metal matrix, different additives, liners and architectures as well as coatings have to be considered. In addition, a selection of adhesives has to be made for use in cryogenic environments. Modern pressure accumulators are made of composite materials (thin inner tanks made of aluminium or polyethylene, reinforced with carbon or glass fibre on the outside). These constructions are much lighter than steel vessels [27].

For the aviation industry, the storage vessels and associated components must withstand defined normal static and failure fatigue load cases as well as the demanding environmental (including shock and vibration) and reliability requirements associated with commercial aviation procedure defined in DO160, DO178 and DO254 [16]. At the same time, the hydrogen storage design and its installation must account for thermal deformation as well as pressure and temperature fluctuations during operation and filling at delta temperatures of ~300 K, when using liquid hydrogen. For most applications, a metal tank gives the best results for the lowest price. Especially in the case of physical storage of compressed hydrogen of up to 700 bar, the storage has to have the form of a pressure tank. When simply operating the vessel, already the thermal expansion of hydrogen leads to higher pressures. Those pressure storages have therefore to be qualified to factors of the design pressure. For the commercial aviation industry, this is regulated by the EASA regulations (CS25 or CS23) [17].

For a high pressure accumulator, the EASA certification regulation (CS25.1435) requests a proof pressure test of three times the design operating pressure and four times for the ultimate pressure. These are the values at which the vessel can deform, but is not allowed to burst. This means that a 700 bar hydrogen accumulator has to be qualified for 2.800 bars in a large commercial aircraft. It is understandable that this design requirement leads to a very heavy unit. On the other side, the design of such pressurized hydrogen storage is very simple as can be seen in Fig. 4. The hydrogen can be filled and extracted by one single pipe. A safety valve or similar system has to be installed and for maintenance for this type of vessel a simply opening for access can be added. The filling status of this type of vessels can be evaluated by measuring the pressure and temperature of the stored hydrogen as well as other system status information by additional sensors (Table 2).

The required design principle of high-pressure hydrogen storages can be overcome using liquid hydrogen. As the storage is at a much lower pressure of maximum 5 bar, the design can be done for 15 bar. Nevertheless, for the operation of a liquid hydrogen (LH2) tank, additional precautions and also design features have to be implemented. For superior gas barrier properties, a liner and tank wall concept with low gas permeability is needed. As the vessel has to store hydrogen safely at cryogenic temperatures for extended durations, with low boil-off quantities, insulation technologies are required that are durable and lightweight. Especially the insulation is a technology for itself. There are different materials used and as well different numbers of layers, even a layer of vacuum is often used, which leads to additional wall sizes with lower quantities for storage and additional weight.

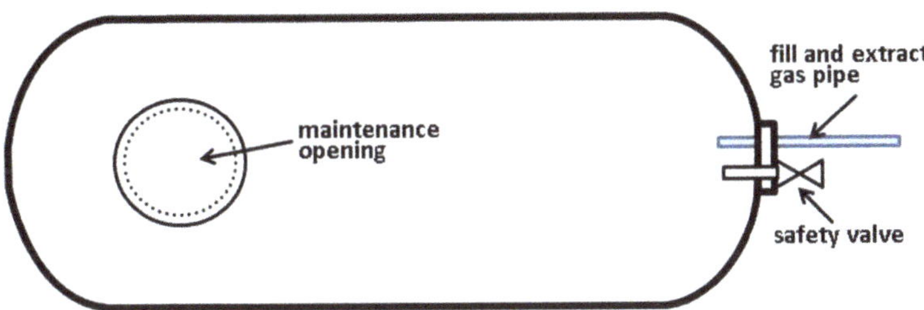


**Fig. 4** Design principle of a compressed hydrogen storage

**Table 2** Pressure qualification values for hydraulic systems (CS 25.1435) [17]

| Element | Proof | Ultimate |
|---|---|---|
| Tubes and fittings | 1.5 | 3.0 |
| High pressure | 3.0 | 4.0 |
| Low pressure | 1.5 | 3.0 |
| Hoses | 2.0 | 4.0 |
| All other elements | 1.5 | 2.0 |

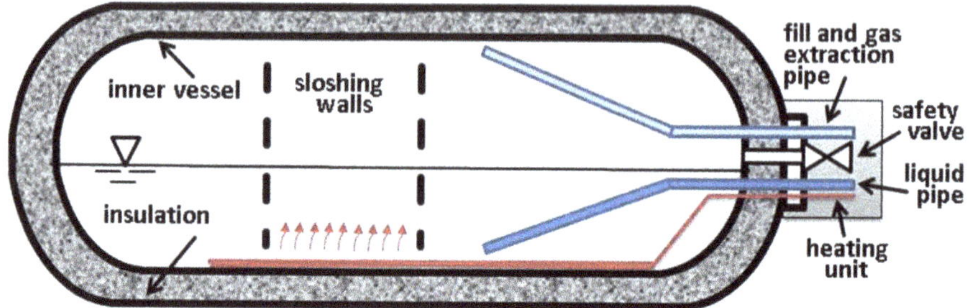


**Fig. 5** Design principle of a liquid hydrogen pressure storage

In Fig. 5, the design principles of liquid hydrogen storages are shown. In addition to the insulation, a heating system is required for the evaporation and warm-up of the liquid hydrogen, because in many cases a supply of liquid hydrogen is not possible, as it needs special equipment. This conditioning system requires a considerable amount of energy. The thermal load has to be injected into the storage with a focus on transient behaviour and start-up procedures, permitting for either centralized boil off or distributed boil off management. For a liquid hydrogen storage, two different pipes are needed. In addition to the gas fill and extraction pipe, a tube for the liquid phase has to be implemented. In many cases, depending also on the size of the accumulator, the moving of the liquid hydrogen has to be taken into account. During the manoeuvres of a large commercial aircraft, this sloshing leads to additional loads on the accumulator wall as well as to fast pressure drops in the vessel due to the mixing of the cold liquid with the warm gas inside. Because of these phenomena, the installation of sloshing walls is used to prevent heavy moving of the liquid hydrogen. This leads to fast settling of the liquid and the extraction of the hydrogen stays stable.

## 4.2 *Storage Control, Monitoring and Indication*

For the operation of hydrogen storages, besides the material selection and functional design definition, other aspects have to be taken into account. The pressure and temperature of hydrogen must be monitored at any time to validate feed and fuel gauging functions under flight accelerations. This hydrogen content control and gauging system are required for every vessel to provide accurate data on mission fuel throughout the flight, minimizing unusable fuel and meeting all applicable airworthiness regulations. To control the flow outside the accumulator, also flow control valves and mass flow metering devices for hydrogen are needed. Adequate safety precautions have to be implemented, which are covered by a secure design together with measurement and monitoring sensors and detection systems. In some cases, also overpressure relief valves and rupture disks are needed to fulfil the safety requirements for certification. The system has to manage both normal and failed system states safely. The tank must also have the means to manage overpressure cases by a venting system safely. Such

**Fig. 6** LH2 tank mock-up made of stainless steel [19]

kind of venting system has to ensure that the hydrogen is released outside the aircraft structure. Otherwise, hydrogen vapour could be ignited by an electrical spark or a lightning strike, and the cryogenic temperatures can harm components in the case of liquid hydrogen spill. Depending on the installation zone of the storage system, also the use of sensors, which can detect hydrogen, is required. The structural integrity of the hydrogen tank can be monitored by structural health monitoring (SHM), which can detect and locate damage in advance. Besides the hydrogen filling status, also the insulation vacuum in the shell has to be monitored for safety reasons by a pressure sensor. A leakage of the vacuum could lead by the thermal expansion of the hydrogen to a fast uncontrolled pressure rise. For this type of liquid hydrogen storage for the aviation industry, a first mock-up in steel was manufactured, shown in Fig. 6 [19].

## 4.3 *Hydrogen Storage Design and Installation*

In the aviation industry, the last extensive research program on the usage of hydrogen was done in the European research project CRYOPLANE around the millennium [20]. In that case, the aircraft was powered by hydrogen instead of kerosene for combustion in the engine. This usage required large hydrogen tanks, which were placed in the roof and therefore led to the well-known unshaped aircraft design. But the idea of that aircraft was again not realized due to technology challenges and decreased interest in alternative propulsion technologies at that time.

In 2010, the idea of using hydrogen in the aviation industry was recapped with the idea to operate a fuel cell. The fuel cell system which was in development was intended to substitute the auxiliary power unit (APU). This installation area was also seen as a safe place as a fire at this place would not lead to a severe damage to the aircraft, like it was also assumed for the APU. The installation of the hydrogen storage in the rear of the aircraft was foreseen in the empennage. The idea was to adapt the development of the storage architecture from the automotive to the aviation industry, because their development was close to serial production. Out of this assumption, the joint research work of Airbus and BMW on the evaluation of cryogenic and liquid hydrogen storage systems was established [18]. Several configurations were investigated using the available space. Figures 7 and 8 show the results of different installation investigations. The design solutions of two companies were used, which were different in size, technology and material. The installation weight was limited to 220 kg for this type of aircraft. This value included 50 kg of hydrogen. The cryogenic compressed storage solution of BMW had a volume of only 8 kg hydrogen, but at a high pressure of 350 bar, therefore, it had to be installed seven times in the bay, which led to the comparison of the highest weight (Fig. 9). By lowering the storage pressure to 100 bar, a

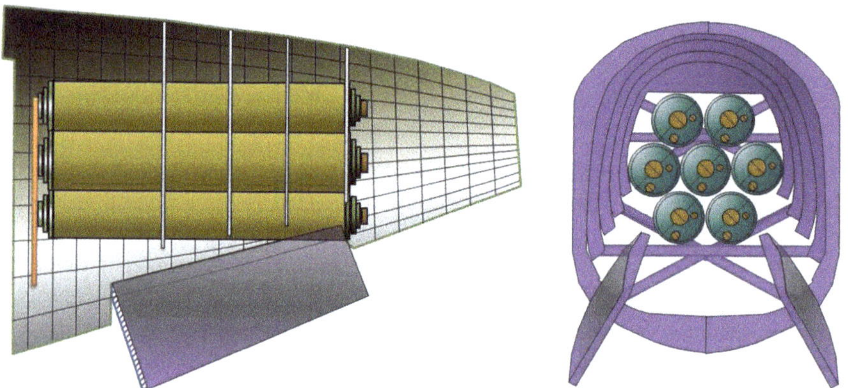

**Fig. 7** Installation of bundle cryogenic compressed $H_2$ ($CcH_2$) tanks in empennage [18]

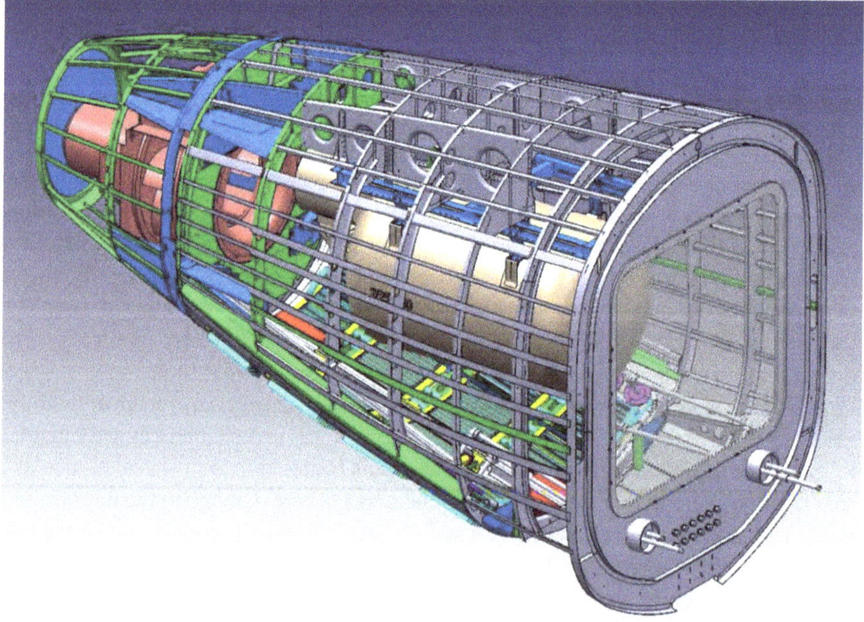

**Fig. 8** Installation of one $LH_2$ tank in the empennage of a mid-sized aircraft [21]

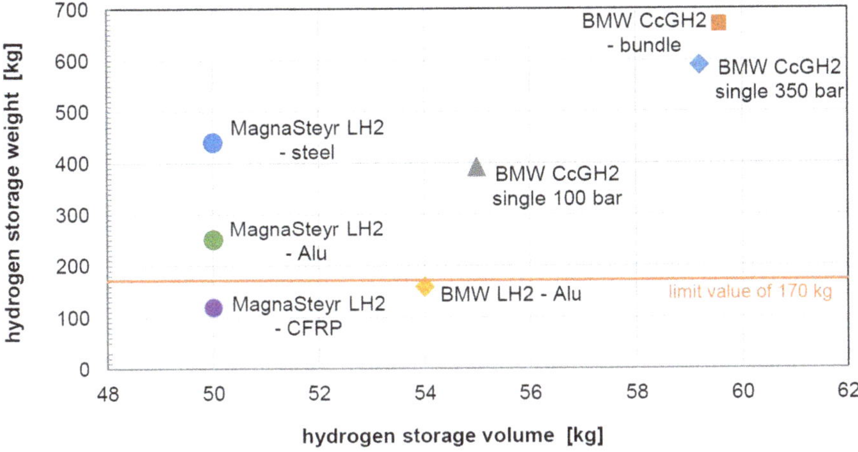


**Fig. 9** Comparison of weight versus hydrogen volume for different tank solutions [18]

lighter version could be designed, due to relaxed design requirements for the storage walls. But the lowest weight was obtained by liquid hydrogen storages made in light weight material of aluminium or carbon fibre. This result underlies why still today liquid hydrogen is in the commercial aviation industry the only choice to store a high quantity of hydrogen with the lowest weight penalty. The gravimetric weight index is used for categorization. The index is defined by the ratio of the hydrogen weight to the sum of the hydrogen and the storage weight, so that 100% is the best value. Taking into account a filled hydrogen storage, an index of 20% is state of the art today. The goal of the ongoing developments on hydrogen storage accumulators is to use different lightweight materials or different material compositions as well as optimized systems to come to an index of 30–40%.

## 4.4 *Tank Venting and Operation Management*

The operation of the storage is of high relevance for the function of a hydrogen system. Depending on the type, compressed or liquid hydrogen storage technology has to be defined, which allows an aircraft refuelling at a high flow rate at any time and which also enables the parking or easy maintenance of the aircraft.

Therefore, for hydrogen storage operational procedures, the focus has to be on:

- Ground and flight operations
- Warm and cold refuelling
- Overnight parking
- Maintenance

For a safe flight of an aircraft, hydrogen in the needed quantity has to be delivered under any circumstances to the propulsion system during the operation of the hydrogen storage. This includes all defined maximal manoeuvres and all operation environments of the aircraft. These requirements imply that a tilting and rotating of the storage and different temperature and pressures have to be taken into account for the design of the storage and its supply system. These facts are the main requirements for aircraft certification. In some cases, this can be ensured by a back-up tank and supply system, with two extraction pipes or pumps. So hydrogen has to be distributed to one or multiple tanks in the aircraft. Nevertheless, this architecture comes along with an additional weight penalty.

It has to be kept in mind that hydrogen as a gas during compression increases in temperature. Therefore, refuelling can cause an increase in pressure and temperature in the tank. This would result in a lower hydrogen volume in the tank at the end of the refuelling procedure for a given fill pressure. This partial fill can be significant for fast refills. Therefore, tests and qualification of all hydrogen components in the temperature range from −260 °C to ambient and under pressure conditions ranging from high vacuum to 1000 bar hydrogen atmosphere are needed. On ground, precooling can lower the temperature and consequently the pressure of the gas going into the tank to maximize the hydrogen filling volume. The adequate filling procedure plays a relevant role in the optimization of the procedure. Therefore, many investigations in the past and today are conducted on the right definition of this process [18].

The integration of a permanently installed tank consequently leads to the need for a filling device. Especially for liquid hydrogen (temperature ~−253 °C), the distance between the refuelling vehicle and the tank should be as short as possible, so that the evaporation losses are minimized. The same applies to the gap between the tank coupling and the tank. Ideally, there is a gradient between clutch and tank on the aircraft side, so that hydrogen flows into the tank even in the event of malfunctions (e.g. failure of the filling pumps). A standard coupling as an interface to refill is needed in alignment with other mobility sectors that allows a safe, reliable operation by the ground staff.

With these restrictions in mind, Airbus investigated the detailed installation of a liquid hydrogen tank in an aircraft [8]. For this design, the needed space of such a system with respect to piping and auxiliary equipment, like the hydrogen distribution system with heat exchangers and pumps, was taken into account as shown in Fig. 10. The major challenge was the integration of the planned liquid-hydrogen tank, a cylinder with approximate dimensions of 1.5 m in length and 1 m in diameter together with its system box that contained the necessary valves and evaporation technologies. The installation space was that of the auxiliary gas turbine (APU) in the empennage, so the last section in the aircraft (point 1).

Point 3 in Fig. 10 indicates the hydrogen piping, which is routed to the consumer system in the aircraft. The pipes must be laid through the pressure bulkhead and at the same time through areas in which other pipes are already located. This results in several challenges, which are the risk of hydrogen leakage by vibrations or mechanical damage. In addition typically aircraft installation requirements apply like segregation of piping and the safe installation in risk areas like tyre burst or rotor failure zones. These aspects must be ensured and validated, having in mind to minimize the pressure loss at lowest possible weight. Due to the mode of operation, gaseous $H_2$ is

**Fig. 10** Installation of a liquid hydrogen tank in an aircraft empennage [22]

required. Ideally the evaporation takes place in the tank. As mentioned before, for the challenges for the operation of a refuelling vehicle an ideal solution for the position of the fuel filler flap on the outer skin above the tank is defined as point 2.

## *4.5 Redundancy and Failure Mode Reconfiguration*

The safe operation of an aircraft and its certification in the commercial aviation industry are challenges, when new systems like the hydrogen propulsion system are designed. Therefore, the willingness of the aviation industry to change or modify an existing and certificated aircraft is low, and the decision is most times driven by business aspects. When going the revolutionary step to introduce a completely new technology like hydrogen propulsion, the risks are even higher. Today, the existing technologies are able to control those risks which are going along with this new energy medium hydrogen. But the rumour around the Zeppelin accident still hinders an open technical discussion.

To recall the facts of this accident, the Diesel was the dangerous fluid, which hinders the rescue. Fluids like today the aviation fluid kerosene do not dilutes and burn away like hydrogen, but as heavy fluid they are floating down and burn on the ground. This situation was visualized in the well-known experiment by the University of Miami, where the ignition of a hydrogen tank and a gasoline tank in a similar car was compared as shown in Fig. 11 [23]. The hydrogen car was safe, because the hydrogen flowed high into the air and burned there, whereas the gasoline car was utterly destroyed, as the gasoline spilled onto the road underneath the car. This behaviour is similar to an aircraft, shown by existing accidents. One good example is the situation where an aircraft caught fire on the ground after the aircraft was parked and the engines were allready off. Because

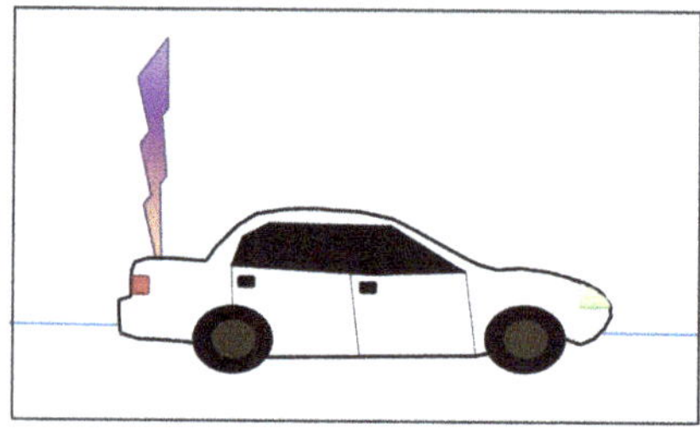

**Fig. 11** Schematic view showing the comparison between the burning of hydrogen (left) and of gasoline (right) from a tank 1 min after ignition [23]

of a leakage in the left-wing tank [24] the kerosene ignited on the hot engines and in a short time the complete aircraft burned down. Therefore today, special trainings are offered by various companies to firer fighting aircraft fires with kerosene [25].

Despite the fact that there are many safety precautions for kerosene, hydrogen is not at all safe. For a safe storage of hydrogen, different aspects have to be considered and a number of risks are identified which need a design solution, such as:

- The results of damage to storage tanks (drop, fire, force, crush, shots, etc.)
- The damage of the insulation for liquid hydrogen vessels leading to vaporization
- The suitability of valves, connections, seals, materials, etc. for cryogen temperature

These risks lead to:

- The release of large amounts of liquid and gaseous hydrogen
- The formation and vaporization of large liquid pools
- The dispersion of large hydrogen clouds

These considerations shall take into account the possibility, that any kind of lightning, for example by a stroke, overheat event or an electrical discharge, causes a catastrophic hydrogen vapour ignition. But also the material could be embrittlement due to the low temperature in case of liquid hydrogen release. The certification paragraphs CS 25.951ff. are especially defining the applicable design precautions for a kerosene fuel system. Nevertheless most of them are also applicable for hydrogen, like the one for venting and lightning protection. Today the working group SAE AE-7AFC and EUROCAE WG-80 defines the applicable rules for hydrogen systems in aviation [26].

For the evaluation of non-fault tolerance for a hydrogen storage system, it should include consideration of structural discrepancies resulting from overstress, ageing, fatigue, wear, manufacturing defects, and accidental and environmental damage. Damage includes conditions that could be reasonably anticipated to occur in the life of an individual airplane due to operation and scheduled and unscheduled maintenance. In addition, potential manufacturing issues in the production process should be considered as probable failures.

The determination of the worst-case failures should be addressed, which can occur in service due to ageing and wear. Depending on the configuration of the aircraft, it is also needed to analyse all cases of a hydrogen tank being hit by some projectile, like a piece of debris from an uncontained engine or tyre failure. If such an analysis would result in a catastrophic scenario, tanks have to be arranged outside of these specified areas. In such a case also the redundancy of the system has to be assessed, because a single failure is not allowed to lead to a complete loss of the hydrogen supply, as an aircraft is not able to land at any place and time.

For the final certification of an aviation hydrogen storage, different kinds of qualification tests have to be performed according to the aviation regulations following the DO160 standard [16], which are:

- Proof and burst pressure tests
- Pressure and thermal cycling tests
- Vacuum (insulation) loss and bonfire
- External impact analysis
- Cryogenic sloshing pressure effect
- Shock and vibration impact

## 5 Concluding Remarks

The appropriate architecture, design and installation of a hydrogen storage are crucial for the reliable operation of a hydrogen system in an aircraft.To ensure this, a proper selection of the storage method depending on the type of aircraft and its operational range and environment has to be done. For a commercial aircraft the usage of liquid hydrogen is the only feasible solution today, having in mind technology maturity, storage size and reachable distances. This need is totally different to the one of a commuter aircraft or an unmanned vehicle. Even for the last categories one can distinguish between large tactical vehicles which have to stay for a long time in the air or the today popular drones, which are in most cases only used for a short time for observation or transportation of goods. In the end, the architect of the flight vehicle has to decide, which hydrogen storage solution has the best value for his device, taking into account cost, quality, capacity, safety and usage time of the storage system.

## References

1. L. Tittel: 1936–1937 LZ 129 "Hindenburg—1937–1987 50 Jahre Unglück von Lakehurst" In "Schriften zur Geschichte der Zeppelin-Luftschiffahrt Nr. 5". Gessler, ISBN 3-926162-55-4, S. 23. 1987
2. J. Klier; M. Rattey; G. Kaiser; M. Klupsch; A. Kade; M. Schneider; R. Herzog: A new cryogenic high-pressure H2 test area: First results. Proceedings of the 12th IIR International Conference: Dresden, Germany, Sept. 2012

3. K. Kunze, O. Kirche: CRYO-COMPRESSED HYDROGEN STORAGE, Cryogenic cluster day, Oxford, Sept. 2012
4. S. Schäfer, S. Maus: Technology Pitch: Subcooled Liquid Hydrogen (sLH2), NOW & CEP Heavy Duty Event, April 2021
5. R. K. Ahluwalia, T. Q. Hua, J-K Peng, S. Lasher, K. McKenney, and J. Sinha: Technical Assessment of Cryo-Compressed Hydrogen Storage Tank Systems for Automotive Applications, Argonne National Laboratory, ANL/09-33, Dec. 2009
6. L. Schlapbach, A. Züttel: Hydrogen-storage materials for mobile applications, Nature 414: pp. 353–358, 2002
7. L. Klebanoff, J. Keller: Final Report for the DOE Metal Hydride Center of Excellence, Sandia National Laboratories, SAND2012-0786, Feb. 2012
8. M. Vogt, L- Röntzsch: Power Paste—Energy Storage Solution, Fraunhofer Gesellschaft, Flyer, also in FORSCHUNG KOMPAKT, Feb 2021
9. K. C. Ott, S. Linehan, F. Lipiecki, C. L. Aardahl: Down Select Report of Chemical Hydrogen Storage Materials, Catalysts, and Spent Fuel Regeneration Processes, Chemical Hydrogen Storage Center of Excellence, FY2008 Second Quarter Milestone Report, May 2008
10. G. Thomas and G. Parks, Potential Roles of Ammonia in a Hydrogen Economy, Feb. 2006
11. M. Hurskainen: Liquid organic hydrogen carriers (LOHC): Concept evaluation and techno-economics. VTT Technical Research Centre of Finland. Report No. VTT-R-00057-19, Dec. 2019
12. R. Ströbel, J. Garche, P.T. Moseley, L. Jörissen, G. Wolf: Hydrogen storage by carbon materials, Journal of Power Sources, 159, pp. 781–801, 2006
13. J. Kleperis, P. Lesnicenoks, L. Grinberga, G. Chikvaidze, J. Klavins: Zeolite as material for hydrogen storage in transport applications, Latvian Journal of Physics and Technical Sciences 50(3), pp. 59–64, June 2013
14. M. S. Turnbull: Hydrogen Storage in Zeolites: Activation of the pore space through incorporation of guest materials, PhD Thesis, University of Birmingham, March 2010
15. A. Züttel: Hydrogen Storage Methods, Naturwissenschaften, 91, pp. 157172, April 2004
16. RTCA DO-160G: Environmental Conditions and Test Procedures for Airborne Equipment; RTCA DO-178C: Software Considerations in Airborne Systems and Equipment Certification; RTCA DO-254: Design Assurance Guidance for Airborne Electronic Hardware
17. EASA (European Aviation Safety Agency): CS25. Certification Specifications for Large Aeroplanes; Amendment 26; CS23: Certification Specifications for Normal-Category Aeroplanes
18. O. Kircher, E. Saefkow, B. Strauß, T. Jordan: CryoSys—Systemvalidierung Kryodruck-Fahrzeugtank, Final Report, Luftfahrtforschungsprogram LUFO IV-4 German research program, final report, May 2012
19. B. Pessl: Innovative Hydrogen Storage Systems A3PS Eco-Mobility. Tech Gate, Vienna, Austria, October 2014
20. A. Westernberger: CRYOPLANE—Liquid Hydrogen Fuelled Aircraft—System Analysis, European funding program FP5-GROWTH, final report, Sept. 2003
21. D. Kastell, ECOCENTS—Effizientes Cooling Center für Flugzeugsysteme, Luftfahrtforschungsprogramm LUFO IV-2, German research program, final report, Feb. 2013
22. D. Kastell: FUCHS—Fuel Cell and Hydrogen System, Luftfahrtforschungsprogramm LUFO IV-4, German research program, final report, April 2017
23. M. Swain: Fuel Leak Simulation, Proceedings of the 2001 DOE Hydrogen Program Review, NREL/CP-570-30535
24. Japan Transport Safety Board: AIRCRAFT ACCIDENT INVESTIGATION REPORT CHINA AIRLINES—B18616; August 2009
25. https://www.draeger.com/en-us_us/Products/Aircraft-fire-training-systems
26. O. Savin: Standardization Activities on Hydrogen & Fuel Cell Technologies for Airborne Applications, DOE's H2@Airports Virtual Workshop, Nov. 2020
27. M. Sippel, A. Kopp: Progress on Advanced Cryo-Tanks Structural Design Achieved in CHATT-Project, ECSSMET—European Conference on Spacecraft Structures, Materials and Environmental Testing, Sept. 2016

# Liquid Hydrogen – Status and Trends as potential Aviation Fuel

Michael Bracha

## 1 Introduction

Liquid hydrogen ($LH_2$) has been the standard fuel in the space industry since the start of the American "Apollo" project, which in 1969 enabled the first manned landing on the moon. Already at that period, NASA investigated possibilities to use $LH_2$ as propellant for civil aviation as well. But only now hydrogen is getting more focus in the aviation industry for the next generation of aircrafts. Two strong drivers are the reason for this development. Driver one is the environmental aspect with climatic change, $CO_2$ emissions, and likewise "peak-oil" which is not in the headlines but can already be felt as, e.g., by the massive decline of the North Sea oil resources. The other one is the unprecedented development of the hydrogen market in the last years which is characterized by strongly growing market applications and by foreseeable falling prices for green hydrogen in the future.

The combination of these two effects has now triggered the aviation industry to start development of hydrogen-powered aircraft. Probably, the further deployment will take place by two different trends. The first one is the field of small aircraft and general aviation, where hydrogen can be used via fuel cells or modified turbines. The first aircrafts of this category are being already tested and could be market-ready in the next years. Hydrogen supply concepts for these aviation applications still have to be developed in parallel and under aviation-specific boundary conditions, but due to the limited hydrogen quantities required for this application, there should be no show-stopper. The second category is large commercial aircraft that dominates the international aviation industry. This aircraft requires much more time for development, and likewise, also the corresponding new $LH_2$ infrastructure is a

M. Bracha (✉)
Michael Bracha - H2 Coaching und Seminare, Munich, Germany
e-mail: info@h2partnermunich.com

 
C. O. Colpan, A. Kovač (eds.), *Fuel Cell and Hydrogen Technologies in Aviation*, Sustainable Aviation, https://doi.org/10.1007/978-3-030-99018-3_2

challenging long-term project. In addition, the ramp-up of the airport infrastructure has to enable a parallel operation of kerosene and hydrogen-operated aircraft.

The main challenge will be the substitution of the enormous quantities of kerosene by green hydrogen consumed in aviation worldwide. Detail engineering will lead to new technologies for aviation, but a good basis therefore should already be available by the presently ongoing infrastructure development for cars, buses, trucks, trains, and—now impending—marine applications. As aviation is an international business, also international coordination and hardware standardization have to follow this development already from an early stage.

## 2 Historical Review: $LH_2$ in Aviation and Industrial Liquefaction

A retrospective view concerning both $LH_2$ in aviation (see Fig. 1) and the development of industrial hydrogen liquefaction demonstrates that important prerequisites already exist for the introduction of this technology in aviation.

### 2.1 $LH_2$ in Aviation

The history of hydrogen in aviation starts with the use of the first hydrogen balloons in 1783 for both unmanned and manned flights. Hydrogen—as a newly used gas—for these balloons was produced at that time by combining sulfuric acid with scrap iron, according Eq. (1):

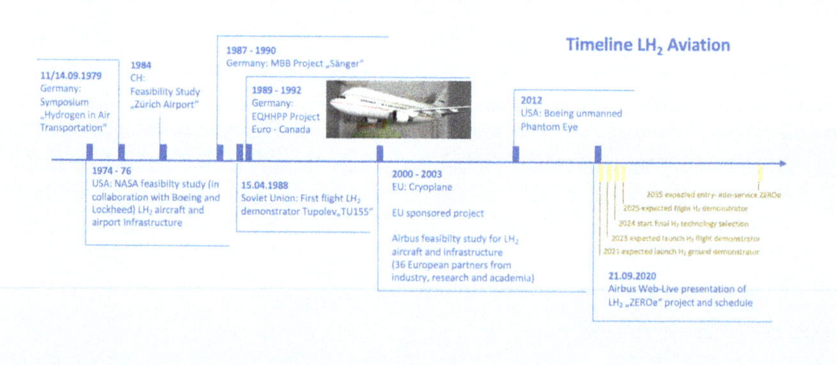


**Fig. 1** Timeline $LH_2$ aviation

$$2Fe + 3H_2SO_4 + 1/2O_2(air) \rightarrow Fe_2(SO_4)_3 + H_2O + 2H_2 \quad (1)$$

The era of the hydrogen balloons was later followed by the era of 'airships' that ended quite abruptly with the famous Hindenburg accident on 6 May 1937 at Lakehurst.

The use of liquid hydrogen in aviation, however, began in 1955/1956 with the American B57-Canberra project. This two-engine demonstrator aircraft was operated with hydrogen in one engine on several test flights, hydrogen being stored in a $LH_2$ tank in one of the wings. These successful tests later initiated an engine test program in the years 1956–1958 at Pratt & Whitney on behalf of the US Air Force [1, 2].

Several years later in 1974–1976, NASA started investigations for use of liquid hydrogen in commercial aviation. While today the driver for the development of hydrogen aviation is the concern of global warming and $CO_2$ emissions, it was at that time the concern of limited availability of fossil fuels (peak oil) when the "oil crisis" took place. Following the awareness that both the development of a new kind of hydrogen aircraft and the build-up of an adequate airport infrastructure would at least require one to two decades, the intention of the NASA study was to start already detailed investigations in order to have the technology available by the turn of the century. Several types of subsonic and hypersonic aircraft were studied. The design of an airport hydrogen infrastructure was contracted as study to Lockheed for San Francisco International Airport (SFO) while Boeing was assigned Chicago O'Hare International Airport (ORD) as reference airport. From today's perspective, the results of these detailed studies still represent a profound engineering basis for an airport infrastructure [3].

In context with the initiative taken by NASA, an international symposium called "Hydrogen in Air Transportation" was held at Stuttgart, Germany in 1979, where experts also discussed on design concepts for $LH_2$ airport facilities and hydrogen aircraft technology. Also, this symposium was concerned about the necessity for an early start of the long-term development program as it was expressed by Willis M. Hawkins from Lockheed Corporation in his introduction with the words: "My hope is that we can somehow wake up our system to be better prepared with alternatives when that inevitable day comes" [4].

Another feasibility study (1982) checked the possibility to install a small $LH_2$ infrastructure at the Zürich airport. Basis for this study was the scenario of one cargo aircraft flight per day with an onsite production capacity of 15–30 t/d (tons per day) $LH_2$. The $LH_2$ production installation was planned to be located between runways Kilo and November for direct refueling the aircraft without use of tank-trucks [5].

Some years later (1987–1990), the former German aviation and space company MBB (Messerschmitt-Bölkow-Blohm) reactivated a project for a space transport system that originally the German engineer Eugen Sänger had conceived. This project officially called "Sänger" was planned as two-stage $LH_2$ system having a scramjet (airbreathing jet engine with combustion taking place in supersonic airflow) in the upper stage enabling hypersonic travel. However, like several similar projects, e.g., the American NASP project (National Aero Space Plan), also the Sänger project was never realized.

Contrary to these ambitious plans for space travel, the former Soviet Union presented a demonstrator aircraft with liquid hydrogen, which had its maiden flight on April 15, 1988. This three-engine aircraft Tupolev Tu-155 (one engine $H_2$ operated) was a special development on the basis of the “workhorse” Tu-154, which was known as one of the most successful Soviet and Russian aircrafts for several decades. Tu-155 was designed for both the operation with liquid hydrogen on board and the operation with LNG (liquid natural gas). LNG was considered as relatively long-term available potential fuel with sufficient sources in the Soviet Union. More than 100 test flights were executed, the majority thereof with LNG (maiden flight January 18, 1989), however, $LH_2$ operation was also successfully demonstrated as feasible for commercial aviation [6, 7].

A feasibility study named EQHHPP (Euro-Quebec Hydro-Hydrogen Pilot Project) was conducted in 1989–1991, which for the first time examined the whole production, supply and application chain of renewable hydrogen produced in Canada and consumed in Europe. Among the various applications, also the concept of a hydrogen powered aircraft (see Fig. 2) was studied by Deutsche Airbus GmbH at that time including later tests for NOx reduction of the $H_2$ turbine [8].

In the following years, several German–Russian initiatives and smaller studies were initiated to continue development on this $LH_2$ aircraft concept; however, a new big approach was then taken within the frame of an EU project called “Cryoplane.” This project started in April 2000 with 35 partners from industry, research and academia from 11 European countries and was coordinated by Airbus Hamburg. All relevant aspects were covered in this comprehensive study, as design of the aircraft, $LH_2$ tank and propulsion technologies, infrastructure issues, environmental

**Fig. 2** EQHHPP project: $LH_2$ aircraft. By courtesy of LBST

compatibility, and safety. The system analysis showed that the concept was viable, no showstoppers were identified. Unfortunately, the next logical step of building up test facilities and a demonstrator aircraft was not taken and the project expired [9].

In July 2010, Boeing's "Phantom Eye" was presented to the public. The Phantom Eye is an unmanned drone with liquid hydrogen tank that benefits from $LH_2$ by high range and especially long operational times of more than 4 days [10].

Meanwhile especially manufacturers of smaller aircraft have started several projects for new or modified aircraft with hydrogen propulsion, some of them have already been flight tested, e.g., ZeroAvia [11, 12]. Common to all these projects is the use of hydrogen as propellant, whereas engine technology and hydrogen storage may vary. For smaller aircraft, fuel cells are often the preferred choice that deliver the electricity for the electrical motors. Hydrogen may be stored as compressed gas or liquid, the tanks itself may be designed as fixed aircraft tanks or as exchangeable modules, where the tank in the aircraft is not refueled but the empty tank module is exchanged by a filled one [13, 14].

On September 21, 2020, Airbus presented ZEROe which is an ambitious program with a clear commitment for liquid hydrogen-based aviation [15]. The program consists of three types of emission-free aircraft, namely a smaller turboprop (<100 passengers), a turbofan (120–200 passengers), and an innovative blended wing body aircraft (approximately 200 passengers). The planned schedule foresees the following milestones:

| | |
|---|---|
| 2021 | Expected launch of hydrogen ground demonstrator |
| 2023 | Expected launch of hydrogen flight demonstrator |
| 2024 | Start of final hydrogen technology selection |
| 2025 | Expected first flight of hydrogen demonstrator |
| 2035 | Estimated entry-into-service of ZEROe |

With the successful implementation of this program, the transition period of taking kerosene aircraft out of operation and replacing them stepwise by hydrogen aircraft finally would have started.

## 2.2 *Industrial Liquefaction*

Figure 3 highlights some of the most important innovations, events, and developments in the history of hydrogen liquefaction.

The history of hydrogen liquefaction started in 1898 when James Dewar succeeded in liquefying hydrogen gas for the first time. The following 50 years was mainly confined to scientific research especially to the investigation of the properties of liquid hydrogen, e.g., the ortho and para configuration (different spin orientations of the electrons in the hydrogen molecule) and other thermodynamic data. Basic work in this context was done at a new laboratory in Boulder, Colorado, operated by National Bureau of Standards (NBS) [1].

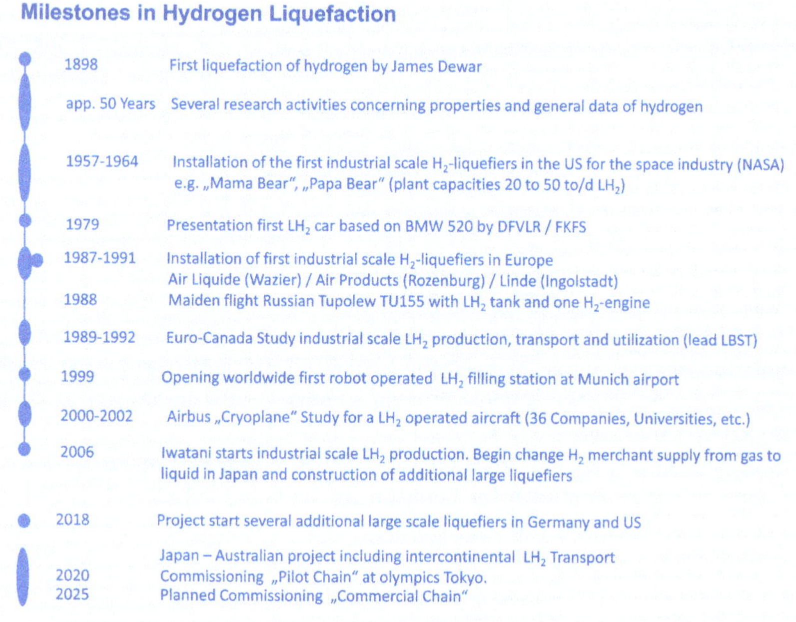


**Fig. 3** Milestones in hydrogen liquefaction

Industrial demand for liquid hydrogen, however, occurred in the 1960s with the ambitious American space program of the "Apollo" project. As propellant for the rockets, large-scale liquefiers were taken into operation for the first time with plant capacities ranging between 20 and 50 t/d. These liquefaction plants in combination with the $LH_2$ storage and distribution systems opened a new chapter in the large-scale $LH_2$ production based on cryogenic engineering technologies that are still common practice today.

Apart from the space applications, the commercial market for liquid hydrogen developed only slowly. However, in the 1980s, three new liquefiers were built in Europe by industrial gas companies in order to supply this emerging market. One of the reasons for this development was the inherent high purity of the liquid product, which was requested especially by the semiconductor industry. Another reason is the logistic advantage of liquid hydrogen compared to pressurized hydrogen. One $LH_2$ truck can typically transport the same hydrogen quantity as about seven trucks with compressed hydrogen (depending on type of $GH_2$ trailer), meaning a substantial cost saving concerning truck and drivers that compensates for the higher liquid price. This effect increases of course with the delivery distance of the transport. As the average delivery distance in the US is significantly higher than in Europe, the percentage of liquid transports compared to gas transports in the US is accordingly also higher.

Presently, the market for liquid hydrogen is rising at an unprecedented scope. Worldwide a large number of new liquefaction plants is under construction or being planned. Especially for mobility applications (cars, trucks, buses, and maritime traffic), a growing market has increased the demand for new liquefaction capacities. Another order of magnitude for liquefaction capacities can be expected when green hydrogen will be used as energy carrier that has to be transported seaborne. Such a project is already in operation between Australia and Japan. Hydrogen is produced and liquefied in Australia and then shipped about 9000 km for consumption to Japan. The present demonstration phase will be followed by a larger commercial phase that shows several technological similarities to the global energy transport of LNG (liquefied natural gas) nowadays [16].

Figure 4 illustrates the development of the global liquefaction capacities. The columns represent the development for 10 years periods; however, the last column confines to a forecast of the increase for the 2020–2023 period only, which indicates the rapid acceleration of the liquefaction market capacities.

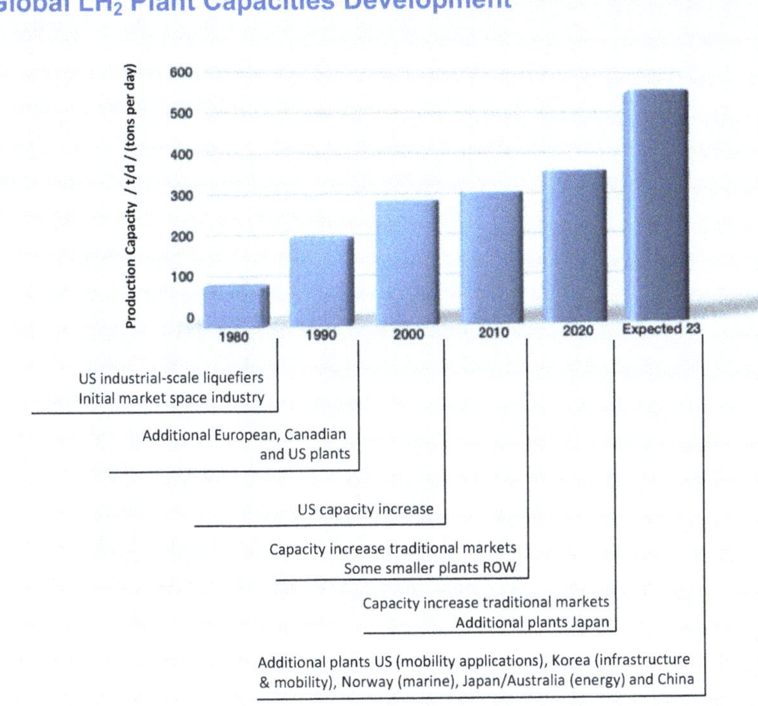


**Fig. 4** Global LH2 plant capacities development

# 3 $LH_2$ as Aviation Fuel

As can be seen in the above-described historical development, liquid hydrogen has always been the preferred aviation fuel for future technology concepts since approximately 50 years. Several factors were the reason for this choice which are explained in the following chapter in more detail.

## *3.1 Density and Weight Considerations*

Hydrogen as aviation fuel enables clean air transportation based on renewable energy and almost emission-free flights with only water vapor as emission. In case hydrogen turbines are used, some NOx deriving from air is also emitted.

Liquid hydrogen as aviation fuel benefits from the high density compared to gaseous hydrogen (see Fig. 5) and the extremely high purity due to its cryogenic production process, which has also a positive effect on the lifetime of the aircraft turbines. For mobility applications, hydrogen pressures of 350 bar for buses and trucks and 700 bar for cars have been established in the market today. Change over

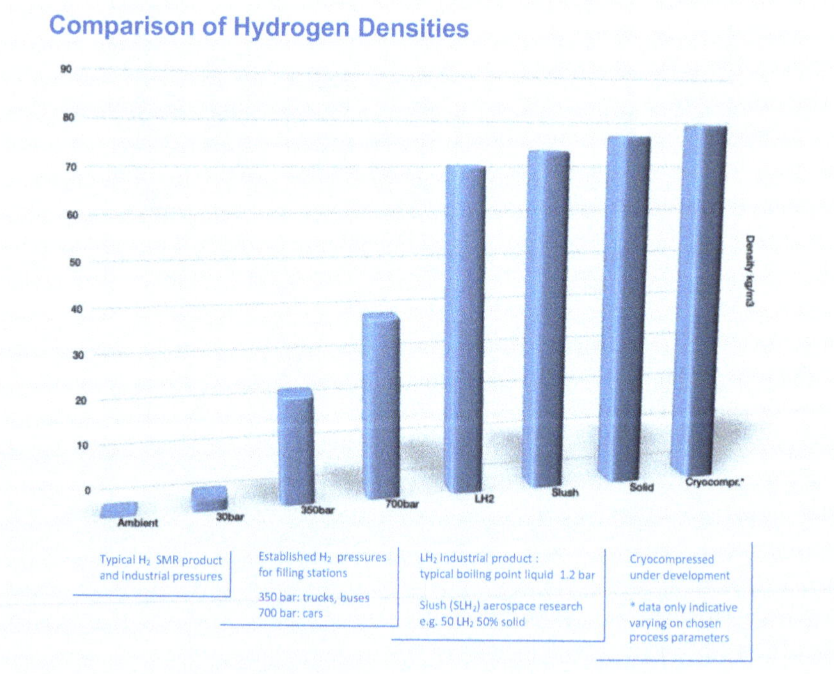


**Fig. 5** Comparison of hydrogen densities

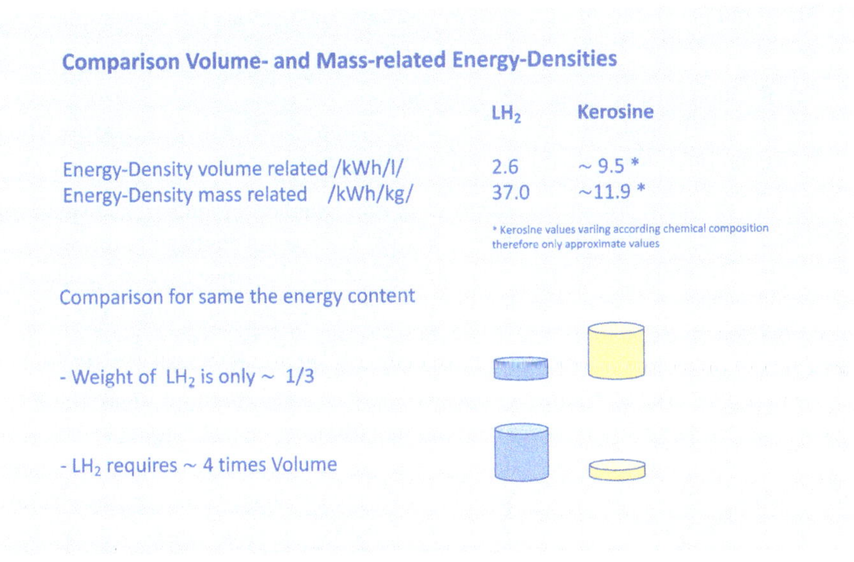

Comparison Volume- and Mass-related Energy-Densities

| | $LH_2$ | Kerosine |
|---|---|---|
| Energy-Density volume related /kWh/l/ | 2.6 | ~ 9.5 * |
| Energy-Density mass related /kWh/kg/ | 37.0 | ~11.9 * |

* Kerosine values variing according chemical composition therefore only approximate values

Comparison for same the energy content

- Weight of $LH_2$ is only ~ 1/3

- $LH_2$ requires ~ 4 times Volume

**Fig. 6** Comparison of energy densities

to liquid hydrogen enables almost a duplication of the density compared to 700 bar storage.

In comparison with kerosene, liquid hydrogen has a three times higher gravimetric energy density but a volumetric energy density of about only one fourth (see Fig. 6). This means that the $LH_2$ fuel itself has a strong weight advantage, which is an important factor in aviation. However, the aircraft empty weight of the $LH_2$ version is generally higher since the tank is heavier than the kerosene storage in the wings, and the larger tank volume also requires a different aircraft design with a larger surface and fuselage volume. The refueled aircraft generally benefit from the higher gravimetric density of hydrogen leading to a lower take-off weight which also has a positive effect of reduced turbine power demand and a lower noise emission. The magnitude of this effect varies between the different aircraft types as regional, short-, mid-, and long-range versions [17, 18].

## 3.2 *Product Development $LH_2$*

Liquid hydrogen, which is a commonly used market product today, can be defined as boiling point liquid at moderate pressure, e.g., 1.3 bar absolute. Between 1.0 and 1.5 bar absolute, the boiling point temperature varies between 20.3 and 21.7 K (data for para-hydrogen). The small overpressure is foreseen as safety measure in order to avoid leakage of other gases, e.g., air, into the product.

The rapid development of hydrogen mobility applications has now initiated the research on two other cryogenic products with similar properties as $LH_2$ [19], namely.

- CryoCompressed Hydrogen (cCH2) and
- Subcooled Hydrogen (sLH2) (The abbreviation "sLH2" is also used for slush hydrogen [mixture of liquid and solid hydrogen])

Both of them are presently subjects of research and development mainly for the hydrogen storage in trucks, but these investigations are not confined to this application only. The cCH2 cycle between tank filling and $H_2$ consumption is operated above the critical pressure of hydrogen (critical point: 33.2 K/13.2 bar) and enables a storage density even above $LH_2$ (depending on selected p/T parameters). A rise of the process pressure from 300 to 450 bar for example would result in a density of about 80 kg/m$^3$. The sLH2 cycle operates at a lower pressure level than the cCH2 cycle with correspondingly lower material requirements on the equipment as piping and tank. The filling procedure of this cycle starts with subcooled $LH_2$. Common advantages of both systems compared to $LH_2$ are the longer holding times before venting and a simpler and quicker fueling process due to one phase flow without cold gas return. Both $cCH_2$ and $sLH_2$ can be easily produced on the basis of $LH_2$.

These practical advantages will probably open the way for a widespread use in the mobility sector. If this is also applicable in the case of aircraft depends on the future overall system detail engineering progress since especially the cCH2 system presently requires heavier tanks due to the elevated operational pressure level.

## 3.3 Potential Future Synergies Based on Heat Sink Potential of $LH_2$

Storing hydrogen in liquid form on board the aircraft opens the way to apply some innovative technologies based on the available heat sink utilization:

- Laminar Flow Control (LFC) of the wings to reduce the skin friction drag: Cooling a surface affects the critical Reynolds number by stabilizing the boundary layer. As a result, the fuel consumption of the aircraft could be significantly reduced, provided the flight mission profile is executed to a large extent at high flight level where ambient moisture will not pose a freezing problem [1, 3].
- Cabin cooling: By heat exchange, the cabin air can be cooled effectively.
- Turbine cooling: The $LH_2$ heat sink can also be used for cooling purposes of the engine and the thermal management, although this benefit has to be balanced against the higher turbine complexity. For supersonic or hypersonic aviation, however, the cooling capacity of the hydrogen fuel would be a substantial advantage due to the high cooling demand of the aircraft friction at the leading edges of the fuselage.
- Superconductivity: This technology is especially well suited for smaller aircraft that are not operated by a hydrogen powered turbine but by a combination of fuel

cell and electrical motor. In this case, a significant power density increase can be achieved by using a superconducting motor instead of a conventional electrical motor.

Another synergy of hydrogen on board an aircraft in general is the possibility to replace the conventional APU (auxiliary power unit) by a fuel cell.

## 3.4 *Safety Aspects*

Safety of hydrogen has been in the focus of discussions since the beginning of its industrial use; therefore, the safety properties of hydrogen shall not be discussed in detail in this chapter [20]. Generally spoken by comparing different fuels, a specific fuel cannot be rated as "safe" or "unsafe," because every specific fuel has its own safety pattern. It therefore makes more sense to define circumstances that are characteristic for a certain application and to evaluate the safety on that basis. For hydrogen aircraft, this scenario would probably be a crash fire hazard [3]. In 1980, NASA performed studies on this basis for crash landings, failed take-offs and ground accidents for the same aircraft type with different fuels including $LH_2$ and jet fuel JP-4.

Following conclusions could be drawn from these studies:

- Tank integrity: In a survivable crash landing or failed take-off the probability of a severe tank damage is lower for a cryogenic tank installed in the fuselage than for a conventional wing tank which is almost unprotected. In addition, a cryogenic tank has a higher structural strength than a conventional aircraft wing tank.
- If liquid hydrogen is spilled, it will evaporate quickly and then dissipate vertically due to its buoyancy in short time, whereas spilled jet fuel will spread beneath the aircraft posing a safety risk to all areas around the spill.
- If a spill catches fire, the duration of the $LH_2$ fire is significantly shorter than for jet fuel (tests with 32 gal resulted in 27 s for $LH_2$ and 7 min for jet fuel). In addition, the thermal energy radiated by the flame is lower for hydrogen than for the jet fuel. These two effects in combination mean that, contrary to jet fuel, a hydrogen accident will probably not heat the fuselage of the aircraft to the point of collapse. Quite similar results were obtained in 2001 by Dr. Michael Swain at the University of Miami [21] who compared the severity of a hydrogen and gasoline fuel leak and ignition of a passenger car. Both tested cars were identical with the exception of the tanks (hydrogen vs. gasoline). After the fire extinguished, the gasoline car was completely destroyed and burnt out, whereas the hydrogen car was almost undamaged, which was also documented in an educational video of the test.

As a conclusion, this means that passengers have a good chance to survive if they stay in the aircraft until the hydrogen is burnt. In contrast passengers of a kerosene aircraft have to be evacuated as fast as possible since the aircraft cabin will probably collapse under the high fire load. For this accident scenario, $LH_2$ has therefore a better safety behavior than jet fuel.

Concerning the airport infrastructure, two safety aspects shall be mentioned. The first one is the possible contamination of the extended distribution system due to the high number of (un)coupling procedures when fueling the aircrafts. If the purging of the coupling is not done properly or by a mechanical defect, a small air quantity could enter the system and accumulate in the course of time by forming locally crystals. Detail engineering of the system has to keep this scenario in mind and foresee proven technical solutions.

The second one is the required safety distance of the $LH_2$ storage tanks to other installations, offices, roads, runways, etc. Unfortunately, for the same setting, RCS (Regulations Codes Standards) varies in a wide range in different countries. It is therefore necessary to pursue an international standardization as far as possible. Furthermore, it has to be considered that not only the safety distance but as well the applicable permitting procedure (and consequently permitting time) depends on the quantity of the stored hydrogen. This may be of interest at the binning of the ramp-up phase when only moderate $LH_2$ quantities have to be stored [22, 23].

## 4 Technology of Large-Scale Hydrogen Liquefaction

For the liquefaction of hydrogen, several processes are available. Some of them are well-known and technically mature while others are subject of investigation in order to improve—especially for larger capacities—the specific energy consumption [24–26].

For the hydrogen liquefaction itself, the process selection depends on the overall economics (CAPEX and OPEX), which has to be optimized for the calculated life cycle of the plant. The process selection is furthermore strongly dependent on the size of the plant. Generally spoken for a small liquefier, e.g., at an institute, a low investment is important, whereas the specific energy demand is not of first concern. For an industrial liquefier with a high liquefaction capacity, however, it is vice versa, low specific energy demand is the driver and main factor in the design of the plant. Increasing the number of expansion turbines in such a plant for example makes sense if the plant efficiency can be improved by this measure. In contrast to fuel cells and electrolyzers which have—within one plant generation—a limited scaling effect on plant size since the capacity is mainly dependent on the exchange surface, a hydrogen liquefier benefits strongly from scaling effects in both CAPEX and OPEX when the capacity is increased.

Figure 7 shows some processes which are presently available for the liquefaction of hydrogen:

For small-scale liquefiers—e.g., in research institutes—hydrogen liquefaction is often linked to an industrial standard helium refrigeration plant that supplies the cooling capacity. Another technically simple method is the high-pressure process where the hydrogen is first compressed to high pressure, then precooled by liquid nitrogen and finally throttled in a so-called Joule–Thomson valve where the liquefaction takes place. The precooling is required due to the inverse Joule–Thomson

$H_2$ Liquefaction Processes (simplified diagram)

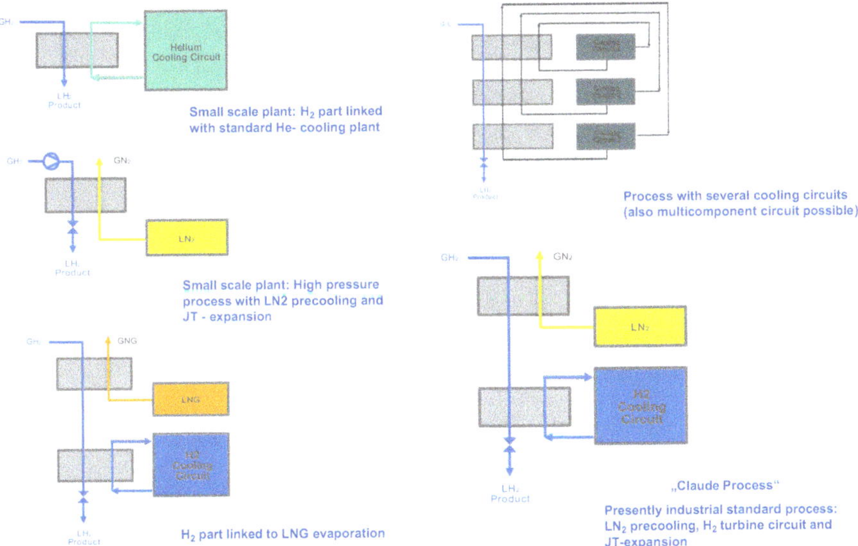


**Fig. 7** $H_2$ liquefaction processes

effect, which would lead to an increase of the hydrogen temperature after throttling above the inversion temperature (pressure dependent) instead of a cool-down below the inversion temperature.

A highly efficient process is the coupling of the hydrogen liquefaction with the LNG (liquid natural gas) evaporation. This process could be realized at large LNG import terminals where LNG has to be gasified anyhow in order to feed the natural gas into the gas-grid. The regasification is nowadays often done, e.g., by burning of NG. By linking it to the hydrogen liquefaction, part of the "cold" energy can be supplied "free-of-charge" and in addition the heating energy for the LNG can be saved.

For large-scale hydrogen liquefaction, several studies and investigations have been carried out in recent years. At the expense of a higher plant complexity, the main target of minimizing the specific energy demand can be achieved thereby. The principle is to use several cooling circuits with different cooling fluids or to use multicomponent refrigerants or a combination of both. Some studies indicate that the specific energy demand for a large plant can be reduced approximately by half by this technology compared to nowadays liquefiers [27]. If these theoretical savings can be achieved in reality has to be checked on a case by case basis. For instance, the use of neon as one proposed refrigerant may result in a theoretical process improvement but is in conflict with the price and availability of neon with respect to the required high quantities in a large industrial liquefier.

The "Claude process" is the most commonly used industrial process (see Fig. 8). Almost all industrial large-scale liquefiers operated worldwide are based on this

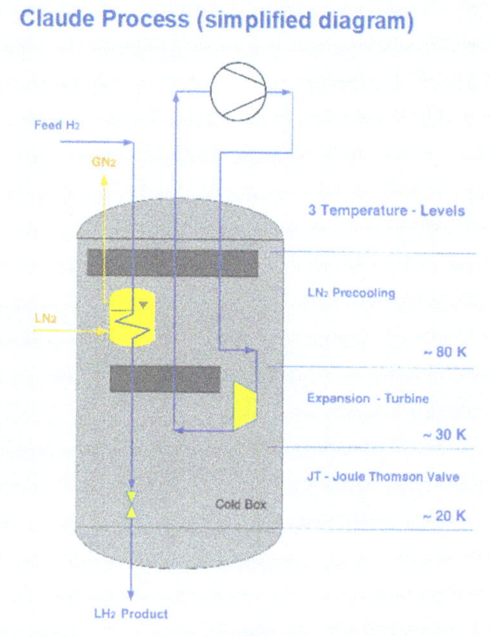


**Fig. 8** "Claude" liquefaction process

technology. The process can be simply described as being operated on three temperature levels. Level 1 is the cool-down of the hydrogen feed-gas by counter flow of evaporating and warming $LN_2$ (liquid nitrogen) to ambient temperature. The second temperature level is a closed hydrogen circuit with hydrogen expansion turbines that deliver the cold for the further cool down of the feed-hydrogen in the temperature range of about 80–30 K. The number and arrangement (parallel or series) of the expansion turbines are subject to the detail engineering of the plant. Below the 30 K level, the final cool-down and liquefaction takes place by throttling the feed-gas in the Joule–Thomson valve before the liquefied hydrogen then is transferred to the storage tank. A very small part of the cold gas after the JT-valve together with the ullage gas (gas above liquid in the tank) at rising tank level is routed back to the pre-compressor of the hydrogen cooling circuit (not shown in simplified diagram).

The hydrogen molecule exists in two spin isomers of the electrons, namely ortho hydrogen (o-$H_2$) with parallel spin configuration and para hydrogen (p-$H_2$) with an anti-parallel spin. At ambient temperature, the normal concentration ratio is 75% o-$H_2$ and 25% p-$H_2$, at liquid temperature the equivalent concentration is almost 100% p-$H_2$. For the conversion of ortho-hydrogen to para-hydrogen special catalysts are installed in the liquefier at different temperature levels (not shown in simplified diagram). This catalytic conversion is done during the liquefaction process in order to receive a stable liquid product, while otherwise the exotherm auto-conversion would take place time-dependent in the stored liquid thus causing partial evaporation of the $LH_2$.

The undesired heat transfer from the ambient to the liquefier equipment is minimized by installing all process parts, e.g., heat exchangers in a so-called cold-box. The cold-box is a cylindrical vessel that is vacuum-insulated and the equipment itself is additionally protected by a multi-layer insulation (MLI) for the reduction of thermal radiation losses.

For reasons of completeness, it shall also be noted that hydrogen liquefaction by a magnetocaloric process was also investigated some years ago; however, this process never gained industrial relevance and remained on R&D status [28].

# 5 Airport $LH_2$ Infrastructure: System and Main Components

## *5.1 Synfuel vs. Hydrogen*

For future aviation, the most likely fuels and propulsion concepts will be:

- Synfuel for conventional aircraft (Remark: In this text, the name "synfuel" has been chosen since it is used commonly in this context (other names: e-kerosene, powerfuel, PtL, P2X, PtX, etc.)
- Hydrogen for turbine-powered aircraft
- Hydrogen for fuel cell-powered aircraft

The main difference between synfuel as a substitute for kerosene and hydrogen as new aircraft fuel is the fact that existing aircraft and fuel infrastructure can still be used with only minor modifications for synfuel, whereas hydrogen requires both a new aircraft type as well as a new infrastructure.

The basic production paths for the two fuels are shown in Fig. 9.

Both the paths have the electrolytic hydrogen production based on green energy as first process step in common. This hydrogen is then liquefied in case of $LH_2$ production or used as feedstock in combination with $CO_2$ for the synfuel synthesis.

A closed cycle and correspondingly a $CO_2$ neutral process is only possible if the $CO_2$ is extracted from air by means of the DAC (direct air capture) process and then emitted again to the air. However, producing a commodity—as a commercial fuel—from air is expensive. This becomes quite obvious when comparing the initial $CO_2$ concentrations of the process. The $CO_2$ concentration in industrial processes, e.g., flue gases, is about 10–20%, whereas the $CO_2$ concentration in air is about 0.04%. Accordingly, this ratio is also reflected in the production costs. Even though the estimation of the DAC production cost is difficult due to the technology development status and other boundary effects, it is presently estimated in the order of approximately 500 €/t which is at least two orders of magnitude above the capture price from industrial gases [29]. In addition, a DAC device requires very large

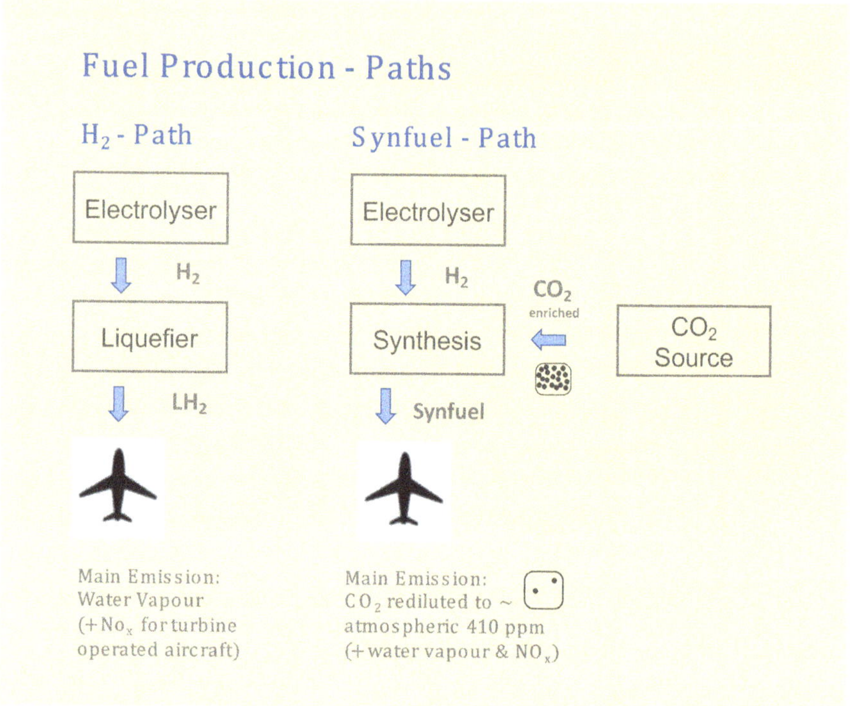


**Fig. 9** Fuel production paths

installation volumes for the $CO_2$ extraction from the air, which—especially with respect to the high amount of fuel consumption at an airport—will typically not be available at most airports or populated areas.

Hence, synfuel synthesis will in most cases be based on an industrial concentrated $CO_2$ source, e.g., from flue gases or cement production, that could otherwise for example be sequestrated or used for greenhouse farming. After use as part of the aircraft fuel the $CO_2$ is then rediluted to air concentration level of about 0.04% again. Such a plant is for instance planned at Heide (north of Hamburg) to supply the Hamburg airport with synfuel [30]. Due to the process prerequisites, the synfuel production will be located outside airports at appropriate industrial sites. Transport and storage of synfuel, however, do not differ significantly from kerosene.

Therefore, synfuel can be considered as an intermediate solution only until new hydrogen aircraft become available on a broader basis and kerosene aircraft in the existing fleet have reached a certain rate of depreciation. During that period of time, a coexistence of synfuel and hydrogen is likely until liquid hydrogen will finally gain an economical break-even point.

## 5.2 The Hydrogen Path

The hydrogen infrastructure on an airport depends on the hydrogen capacities needed, respectively, the airport size (small airport/hub, etc.) and on the percentage of hydrogen flights which will continuously grow with time. In the initial phase with only some selected flights, the hydrogen infrastructure requirements will be significantly different from those when already a certain percentage of hydrogen aircraft is in operation or even when the conversion of the fleet is finished.

Due to the relatively large amounts of $LH_2$ that are required for a basic hydrogen aircraft traffic at a larger airport, external supply is a useful solution for the initial phase of the transition process only or for very small airports with mainly general aviation. Container transport or transport via rail is the main supply option in this case. Direct $LH_2$ transfer to the airport via vacuum insulated pipelines is limited to shorter distances since the losses are directly proportional to the pipeline length. A special case would be an airport directly located at the coast in combination with a commercial $LH_2$ import chain via ship as the for instance the Australian–Japan project Hystra. In this case the airport storage tanks could be filled directly from the transport ship [31].

The following concept description, however, will be based on the assumption that already a certain percentage of the flights will be operated with hydrogen at a larger airport. This means that the liquefaction of the hydrogen has to be arranged at the airport, whereas the production of the gaseous hydrogen may be done externally (hydrogen pipeline to the airport) or internally (electricity import for the onsite electrolyzers).

## 5.3 Basic Concept and Main Units

Depending on the size of the airport and the required $LH_2$ capacities, a variety of different design concepts for an airport hydrogen infrastructure are feasible. Also, site-specific utilities and boundary conditions have an influence on the process design. Therefore, for each airport a study for the optimum concept has to be done in advance. The main difference between an airport of today and a larger future hydrogen-based airport is the fact that the fuel for today's airport is delivered to the airport, whereas the liquid hydrogen has to be produced at the airport. For larger demand capacities neither long-distance cryogenic pipeline transport from an external source nor trucked-in $LH_2$ with a high number of daily trailer transports are reasonable solutions.

The following concept was selected as typical example for a larger airport when already a certain percentage of the total traffic volume is based on hydrogen flights. A similar system had been described by G.D. Brewer in 1976 in a detailed airport infrastructure study [32]. San Francisco International Airport was chosen as reference airport at that time with an onsite liquefaction capacity of 1000 t/d (tons per

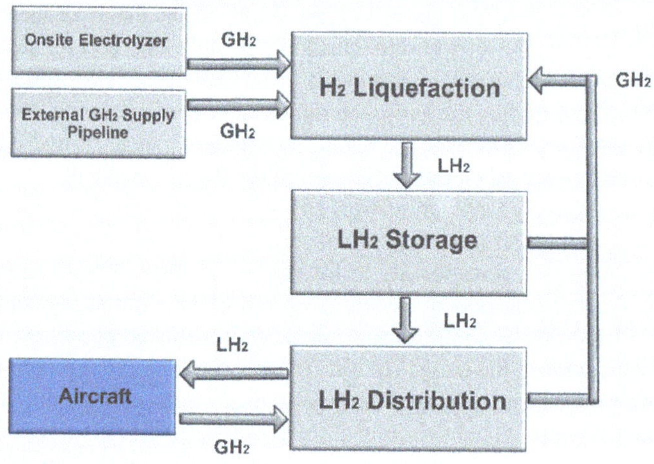


**Fig. 10** Concept of airport hydrogen infrastructure

day) subdivided into four liquefaction units with 250 t/d each, and special emphasis was given to the recovery of the operational boil-off "losses."

The basic concept of such an airport hydrogen infrastructure consists of the following main units (see Fig. 10):

- $GH_2$ Supply: Onsite electrolyzer or $H_2$ pipeline
- Hydrogen liquefier and recondensation
- $LH_2$ storage
- $LH_2$ distribution system (cryogenic transfer lines)
- $LH_2$ fueling station for the aircraft
- Utilities and service installations

## *5.4 $GH_2$ Supply: Onsite Electrolyzer vs. $H_2$ Pipeline*

While it makes sense to liquefy larger quantities of hydrogen directly at the airport, the gaseous feed-hydrogen can be produced onsite at the airport or it can be provided via pipeline to the airport.

**Onsite $GH_2$ production:** A complete production chain based on external electricity supply for electrolysis and subsequent liquefaction is one option for larger $LH_2$ quantities. It is, however, the option requiring the highest ground area demand. As the typical ground area demand for the electrolysis for a chosen capacity is larger than the corresponding area required for the liquefier, it could therefore become a bottleneck for existing airports in a densely populated area. In addition, the water demand for the large-scale $GH_2$ production should also not be underestimated especially for airports as for example in desert environments like Saudi Arabia. It should be noted that nowadays the largest part of the worldwide hydrogen production is still based on the fossil steam reforming process [33]. However, for a future oriented technology such as hydrogen aviation, the steam reformer will probably lack public acceptance, and "green" hydrogen will be demanded at least for new production plants. Therefore, the steam reforming process is not considered in this context.

**Pipeline Connection:** Probably the most preferred future solution, however, is the hydrogen supply via pipeline in combination with an onsite liquefaction plant. In this case, precious ground area can be saved and such a grid connection should guarantee a reliable and cost-efficient hydrogen supply. A long-distance pipeline connection has typically also sufficient buffer volume to tolerate short-time trips of the production plant without jeopardizing the consumer's supply reliability. Several large projects are presently in the planning phase for a basic and later extended hydrogen pipeline grid in Europe either by installing new pipelines or by retrofitting natural gas pipelines to hydrogen operation. The development progress of global green-hydrogen projects is impressive; just within half a year (December 2020 to June 2021), the number of gigawatt-scale projects worldwide increased from 13 (50 GW) to 24 (207 GW) [34]. This trend will probably continue, and consequently, with the growing number of projects and growing capacities, the economics of scale will bring the specific hydrogen price continuously down. But not only the growing number of applications but also the technology itself is progressing. A new process is being developed for instance at the German "AquaVentus" initiative. Being basically an offshore windfarm initiative with many subprojects along the value chain, the generated electricity is not delivered to shore via an electrical transmission cable, but the electricity is transformed offshore into hydrogen either at the wind turbine itself or on an electrolyzer platform and then further transported in an undersea pipeline named "AquaDuctus" (see Figs. 11 and 12). The AquaVentus initiative is designed to produce at its final stage of extension approximately one million tons of green hydrogen annually, which is the equivalent quantity to operate for example eight million FC cars [35]. Jimmie Langham, managing director of AquaVentus, explains: "The pipeline concept has a lot of advantages. Transporting energy via molecules in a pipeline instead of electricity transport saves expenses for five high voltage sea-cables, so-called HVDC (High Voltage Direct Current transmission). Likewise, the pipeline can act as storage buffer for the hydrogen. Crosslinking to wind farms in the Netherlands and Denmark is also possible. AquaDuctus can become a starting point for a large German and northwestern European offshore-hydrogen-hub and can contribute substantially to the low-cost transport of green hydrogen on an industrial-scale" [36]. Considering the growing potential of this kind of hydrogen supply, also the German "Nationaler Wasserstoffrat" has recently recommended to connect airports to the planned hydrogen grid [37].

**Fig. 11** Wind turbine with integrated $H_2$ production. By courtesy of AquaVentus

**Fig. 12** Offshore platform for $H_2$ production. By courtesy of AquaVentus

## *5.5 Hydrogen Liquefier and Recondensation*

The hydrogen liquefier is the central unit of an airport $LH_2$ infrastructure (see Fig. 13). At least two redundant units should be installed to guarantee a high product availability.

High-purity hydrogen and electricity have to be supplied to the liquefier and $LH_2$ as product is transferred to the storage tanks. Depending on the quality of the feed

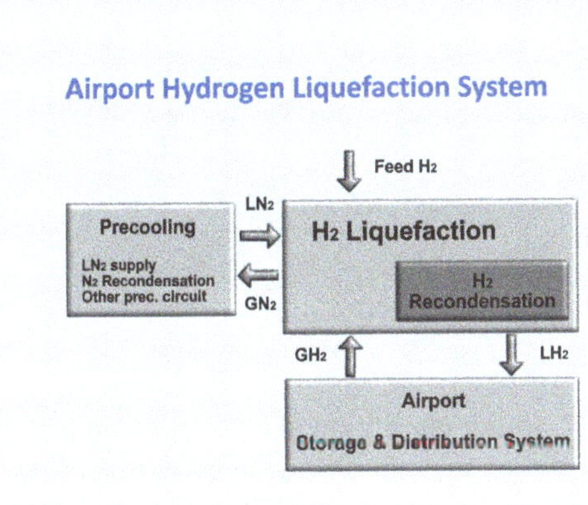


**Fig. 13** Airport hydrogen liquefaction system

gas, a purification unit has to be foreseen additionally if the $GH_2$ quality does not fulfill the liquefier's inlet specification.

Depending on the chosen process, the liquefier has typically also to be supplied with liquid nitrogen for the precooling. This is presently done by many liquefiers via an external $LN_2$ supply, but with growing plant capacities, a closed condensation loop for nitrogen is preferable and more economical.

The collected hydrogen cold gas of the infrastructure distribution system should be returned to the liquefier itself if the installation is designed for this application. However, for a better flexibility and stable liquefier operation, a separate recondensation unit that can also be operated independently should be favored. Apart from economic considerations, collection and recondensation of cold gas is also an important safety factor at an airport as operational venting of hydrogen is avoided.

## 5.6 *$LH_2$ Storage*

Comparing the volumetric energy densities of $LH_2$ and kerosene, it is obvious that the $LH_2$ storage volume has to be increased approximately by a factor of 4 in order to store the same energy as in the kerosene case. But as hydrogen is liquefied at the airport—whereas kerosene is supplied to the airport—the average storage time and accordingly the storage volume can be reduced. If the hydrogen is liquefied at high plant reliability (e.g., redundant units), this allows to fuel aircraft also indirectly from the liquefier outlet with the onsite tanks functioning as buffer volumes for short-time fluctuations instead of long-time storage volumes. This relationship of the stored and continuously produced hydrogen for direct use has to be balanced for each airport individually depending on the onsite boundary limitations and logistic considerations.

**LH₂ Tank-Sizes**

| Type | Standard | Single-unit | Single-unit (on-site fabrication) |
|---|---|---|---|
| typical size geometrical /m³/ | ~ 5 - 60 | < 300 | e.g. 3.800 |
| typical „boil-off loss" /%/d/ | ~ 0.6 – 1.4 | ~0.2 | ~ 0.03 |
| typical design | cylindrical | cylindrical | spherical |

**Fig. 14** $LH_2$ tank sizes

In addition, the $LH_2$ storage capacity will be ramped-up with the introduction of hydrogen flights, whereas on the other side, the kerosene storage requirements can be ramped down accordingly in the long run. This effect will help to limit the necessity of new storage area at the airport to a certain extent.

Technology for the storage of liquid hydrogen is proven, tanks available vary from some $m^3$ geometrical volume to about 3800 $m^3$ at Cape Canaveral for space applications (see Fig. 14). Due to transport limitations $LH_2$ tanks can only be delivered up to approximately 300 $m^3$ volume from a manufacturer site, larger storage sizes have to be built onsite (spherical tanks). Specific boil-off losses strongly decrease with rising storage capacity of the tank. If the tanks are integrated into the airport cold gas system, then no "boil-off" hydrogen is lost and the "energy loss" is limited to the recondensation energy of the cold gas, which is lower than the liquefaction energy of ambient temperature hydrogen gas.

## 5.7 *$LH_2$ Distribution System*

When handling of $LH_2$ at an airport has exceeded a certain capacity threshold, one adequate technical solution is a closed $LH_2$ and cold gas return circuit connecting liquefier resp. recondensation system, storage system, and aircraft refueling connections. The advantages of such a system are obvious, namely safe handling of operational hydrogen cold gas during aircraft filling, reduction of cool-down losses and cool-down times with corresponding reduction of thermal material stress and economical process optimization by cold gas reutilization.

Cryogenic transfer lines of a $LH_2$ distribution system are designed according the same principle as cryogenic tanks, having an inner line which is surrounded by an outer line. The volume between the two lines is evacuated, and additionally, the inner line is protected by a multilayer insulation (MLI) against losses by heat radiation. This principle allows a strong reduction of the heat leak from the ambient to the liquid hydrogen. Nevertheless—especially with respect to the total length of an airport transfer line—a certain heat leakage is unavoidable. Therefore, a small constant flow in a closed-circuit system shall permanently maintain cryogenic operating conditions that allow quick start-up of aircraft filling with reduced cool-down times when required at the hydrant.

Cryogenic transfer lines have already been realized also for longer distances. At Cape Canaveral or at rocket test installations, $LH_2$ lines of several 100 m distances are commonly used. At the international research center for particle physics "CERN" at Geneva, even an underground cryogenic line for liquid helium (temperature below $LH_2$ level) with a circumferential length of 27 km has been in operation for the cooling of the superconducting magnets since decades [38]. Technology for the design of a cryogenic airport distribution system is therefore available; logistical and economic considerations will be the main design criteria. The liquefier resp. recondensation unit has to be designed to cover also the additional capacity which corresponds to the insulation losses of the distribution system minimizing the necessity of operational system cool-down procedures.

## 5.8 *Fueling of the Aircraft*

Apart from filling the aircraft tank with liquid hydrogen, the filling with cryocompressed subcooled hydrogen or the exchange of the empty hydrogen tank by a full tank could be considered as alternatives.

The tank exchange is an interesting solution for smaller aircraft where this kind of procedure is feasible because it avoids the installation of an airport $LH_2$ infrastructure by delivering the filled hydrogen storage pods to the airport and returning the empty units to the hydrogen production and filling station (see Fig. 15). This concept—including a retrofit of the aircraft to hydrogen operation—is presently investigated by "Universal Hydrogen" for the regional turboprop aircraft segment [39, 40].

As already mentioned in Chap. 2, it is not yet clear, if the use of cryocompressed or subcooled hydrogen for aircraft is a viable alternative due to the higher tank requirements (pressure design), meaning higher tank weight. Using one of these fuels, however, would simplify the filling procedure since no cold gas return would be required.

In the following description, however, it is assumed that the aircraft will be refilled with liquid hydrogen from the distribution grid at its parking position, which is comparable to today's kerosene filling. In the initial phase of the infrastructure development, a refilling at a special filling position could also make sense. In this case, investment cost could be reduced, but the turn-around times per flight would be higher.

The filling procedure itself still has to be defined and optimized for the large aircraft tanks especially with respect to the limitations of the turn-around time [41]. In case of cold gas generation, during the filling procedure or during refueling of a warm system after planned maintenance, a safe gas routing is necessary. Today air in the kerosene tanks is vented at the aircraft to the atmosphere during filling. In case of hydrogen, this procedure (ullage hydrogen gas) would cause a safety

**Fig. 15** Aircraft hydrogen tank exchange. By courtesy of Universal Hydrogen

problem for the environment of the aircraft. In addition, the density of cryogenic hydrogen gas is comparable to the density of air meaning that a cold gas cloud can shortly spread horizontally until the buoyant effect takes place when the gas is warming up. Therefore, the only reasonable handling is a controlled routing in a cold gas loop to a flare system or preferably to a recondensation unit for reutilization of the cold gas and for minimizing losses. Also ground overnight parking or long-time overhaul with cold tanks necessitate a safe connection of the aircraft with the airport's hydrogen infrastructure.

The connection of the aircraft with the filling coupling before the filling process can start is more complex and requires more procedures for $LH_2$ filling than in the case of kerosene. While for conventional filling, a good and tight mechanical connection is generally sufficient, the cryogenic coupling has to ensure first a removal of air (evacuation step), proper purging (typically with helium), and cool-down to $LH_2$ temperature until the hose is filled with liquid hydrogen and ready for filling.

For an advanced $LH_2$ aircraft filling system, the application of an automated, robot-operated system connected to the fueling hydrant should be considered. Such a technology can ensure an uninterrupted sequence that may partially compensate for the additional time requirement due to the higher number of process steps compared to kerosene filling within the turn-around time. The filling coupling should be standardized internationally, maybe in combination with similar marine fueling applications if the design flow rates are comparable. The hydrogen filling procedure itself at today's gaseous vehicle hydrogen filling stations is already highly automated and surveilled by checking various process parameters (e.g., pressure test for tightness). An additional automated docking of the couplings for liquid hydrogen would therefore be a next logical step increasing the safety of the process by automatically verifying $LH_2$ connection tightness and correct purging procedure before the sequence step "start filling" is unblocked.

At the Munich airport, such a system had already been taken into operation for the worldwide first $LH_2$ robot filling of the BMW 7-series in May 1999 (see Figs. 16 and 17). The BMW 7-series "clean energy" project prototypes had been equipped with a LH2 storage tank that supplied the hydrogen to the 12-cylinder combustion engine. Main design criteria for this kind of public $LH_2$ filling were safety and maximum comfort for the driver. The driver did not even have to leave the car, via access card he activated the system from the car window, and the robot executed the docking procedure, refueling process, and final undocking with a four-axis kinematics [42].

## 5.9 *Utilities and Service Installations*

Apart from the main units described above, additional installations have to be foreseen for the airport hydrogen infrastructure:

**Fig. 16** Opening of filler flap before connecting $LH_2$ coupling. By courtesy of BMW Group

**Fig. 17** Activating $LH_2$ robot filling by driver from car inside. By courtesy of BMW Group

- Electrical systems: Connection to the electrical grid and transformers to supply the required voltage level, e.g., for the compressors of the liquefier
- Safety system for controlled flaring (hot flare avoiding direct hydrogen release), distributed points of hydrogen detection, and alarm systems
- Control room and maintenance facilities
- Analytical system for surveillance of $LH_2$ product quality and potential impurities in the distribution and filling system at different locations
- Tank conditioning system for aircraft service and inspection. When an aircraft has to undergo regular service checks or troubleshooting activities also the tank has to be accessible. The cold, partially filled tank has to be emptied of hydrogen, inerted, warmed up, and put under an atmosphere ready for intervention. After inspection—in reverse sequence—the tank has to be cooled down and inerted and finally filled with liquid. For this procedure, a special installation has to be foreseen.

These installations have to be considered already in the early phase of ground area planning for the airport hydrogen infrastructure.

# 6 Challenges and Project Management for Long-Term Airport Hydrogen Phase-In

## *6.1 Airport as Hydrogen Hub*

Since the number of new airports that will be built in the future and that could theoretically be designed already as "hydrogen airports" from the beginning will be relatively small compared to existing airports that have to be adapted to the phasing-in of hydrogen aircraft, main emphasis has to be given to such a long-term development. The foreseeable phase-out of kerosene and mid- to long-term phase-in of hydrogen as aircraft fuel require a new approach in the transformation of existing airport fuel infrastructures. Installation of a hydrogen infrastructure on a "green field" basis could be accomplished in some years, however, existing airports have to ramp up the hydrogen infrastructure in parallel with the capacity demands of hydrogen flights while still maintaining fuel service for the present kerosene and probably near future synfuel flights. All this has to be done in an economically well-balanced manner, because the installation of large hydrogen capacities that will be fully utilized only several years later would pose a critical financial burden that most airports can probably not finance. In addition, in most cases, the required space for additional installations would not be available but could be created by a stepwise dismantling of the kerosene infrastructure.

In a first step, some airports are already trying to install a "non-aviation" hydrogen infrastructure as hub for different $H_2$ applications. For this "kick-off" situation, all kinds of possible synergies are used, which is a logical and necessary starting point. Accordingly, mobile hydrogen applications shall be transferred to hydrogen operation. At a large airport, several mobile applications are ideal candidates, e.g., public taxis, GSE (ground support equipment), airport material handling, and freight logistics for trucks and trains. The airports of Paris and Groningen have already started with the installation of such a "non-aviation" basic $H_2$-infrastructure [43–45]. Synergies concerning the demand for heating and cooling of buildings at a large airport can also be achieved by integration into the hydrogen infrastructure.

This first step will also trigger the necessity of gaseous hydrogen production at or close to the airport. In a later phase, the installation of a liquefier could initially serve, e.g., the transport $LH_2$ truck logistic and the growing $LH_2$ merchant market supply of the industrial gas industry as stable baseload for a new plant. This concept could then guarantee $LH_2$ delivery for the first $LH_2$ flights as soon as the aircrafts become available.

## 6.2 *Technology Development*

The basic technologies needed for an airport infrastructure, e.g., liquefier or cryogenic transfer lines are already available. Not available so far are special filling installations (e.g., robot-assisted flexible transfer lines for the aircraft connection) and the coupling itself for the aircraft. These have to be developed, and an international cooperation for regulations, codes, and standards is mandatory. Also process parameters for filling and storage or para-content of the $LH_2$ have to be optimized from an economical point of view for aviation applications.

Since every airport has a unique design, the step-wise build-up of a hydrogen infrastructure will have to be planned individually. However, for reasons of economics, it is also useful to standardize engineering and unit production as far as possible. Therefore, a distinction between "site-specific" and "site-independent" is proposed as basis for the future airport project management. "Site-independent" means for instance design of a certain $LH_2$ capacity production plant that can be installed at several airports. "Site-specific" in this context means that connection to the onsite electrical supply or the $LH_2$ distribution system of this plant has to be adapted to the local requirements. Likewise, for instance, also the permitting process could have a standardized part concerning the liquefier as basic technical unit and a specific part considering the integration of the liquefier into the individual airport location.

### 6.3 Project Management

Conventional project management is responsible to execute the realization of a predefined project based on a technical specification until the plant has been commissioned and can be handed over to the client [46]. As the process of transition from kerosene to hydrogen at an airport will be a long-term process with several intermediate stages, not only the intermediate stages itself will be subject for the project management but also the long-term perspective and the coordinated planning with other airports. Such a coordinated planning makes not only sense to handle the prerequisites for point to point flights but also to find synergies for the hardware development. This may be illustrated by the following example: If, from the initial phase, the hydrogen demand at a hub is constantly growing, a new production plant has to be installed in parallel with sufficient buffer capacity for further growth. The original low capacity plant could then be relocated to a smaller airport that has just begun to introduce hydrogen aviation. This kind of strategic planning would have taken the relocation or a planned plant revamp already into consideration in the design specification of the installation. Such a coordinated (national and international) airport alliance would be beneficial and economically superior to the individual transformation planning of each airport.

Therefore, a long-term planning is mandatory which requires a certain flexibility with respect to the required fuel capacity demands by taking out of operation older aircraft and replacing them stepwise by hydrogen aircraft (see Fig. 18). Accordingly, the long-term planning and project management have to be organized in a manner that will significantly differ from typical project executions models presently in practice.

## 7 Concluding Remarks

In the last years, hydrogen energy applications worldwide have significantly grown in size and numbers, which can be considered as disruptive change toward a new energy system. Japan, China, Korea, Europe, and US can be considered as lead regions, mobility applications, and broader industrial use are the first sectors of this transformation.

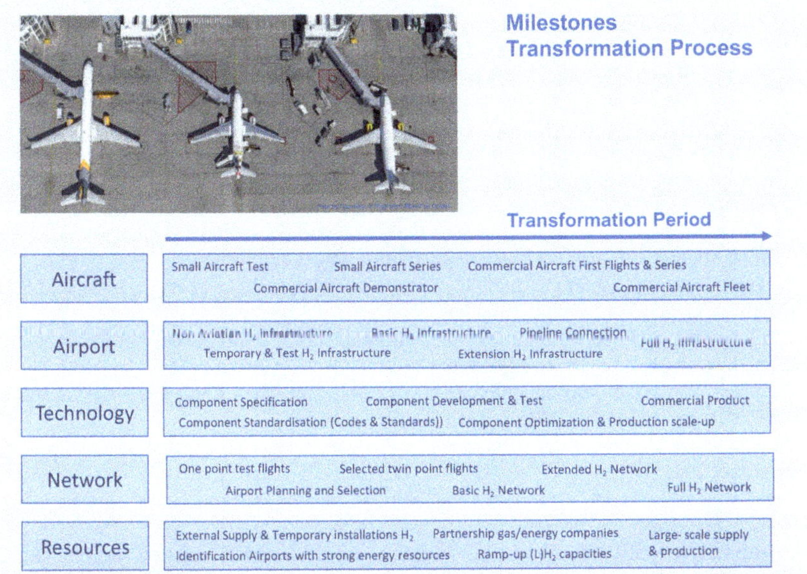


**Fig. 18** Milestone transformation process

The aviation industry is now challenged in two respects. First, the corona situation has led to an economical breakdown while at the same time a switch over to hydrogen aviation will request enormous investments in new aircraft and infrastructure. And it will be a transformation that will typically last longer compared to, e.g., transformation of trucks or buses as both development times and life cycle times for aircraft are in another order of magnitude, probably decades.

Due to this longtime switch over process, new concepts for the installation of hydrogen capacities have to be developed since an existing kerosene-based airport will not be replaced by a newly built hydrogen airport on short notice, but it will be a slowly growing process where both the kinds of aircraft will have to coexist for a certain period of time. This underlines even more the importance of strategic planning, high flexibility, continuous integration of technical development, and an infrastructure project management that is not oriented on one and only final solution but on a "moving target." A new era has begun.

## References

1. W. Peschka, "Flüssiger Wasserstoff als Energieträger", Springer Verlag Wien—New York, 1984
2. Pictures: Copyright of Archiv der Luftschiffbau Zeppelin GmbH, Friedrichshafen, FRG.
3. G. Daniel Brewer, "Hydrogen Aircraft Technology", CRC Press, 1991, 5838, ISBN 0-8493-5838-8

4. DGLR/DFVLR International Symposium "Hydrogen in Air Transportation" September 11–14, 1974, Stuttgart, Germany
5. H.P. Alder "Hydrogen in Air Transportation. Feasibility Study for Zürich Airport, Switzerland", International Journal of Hydrogen Energy, Vol. 12, No. 8, pp. 571–585, 1987
6. P. Zerger, Flug Revue, "Wie Tupolew mit der Tu-155 Geschichte schrieb", 26.09.2020
7. AeroBuzz.Fr, "Le 15 avril 1988, le Tu-155 vole a l'hydrogene", 2.10.2020
8. R. Wurster, Chapter 4 "Die Verdienste von Ludwig Bölkow zur Etablierung von erneuerbaren Energien und Wasserstoff in Deutschland", page 36 ff. "Von der Vision zur Realität—Festschrift zum 100. Geburtstag von Ludwig Bölkow", ISBN 978-3-929371-27-7
9. H. Klug, R. Faass, "Cryoplane: hydrogen fueled aircraft—status and Challenges", Air & Space Europe Volume 3, Issues 3–4, May–August 2001, Pages 252–254
10. Aviation Today, G. Guarino, "Boeing's Phantom Eye Flies Higher, Longer in Second Flight", February 27, 2013
11. E&T Engineering and Technology, E&T editorial staff, "Commercially available hydrogen plane takes flight", September 25, 2020
12. Deutsche Wasserstoffvollversammlung 27.01.2021, Prof. Dr. J.Kallo EWS Universität Ulm/DLR Stuttgart, "Go4Hy2—Fliegen mit H2/BZ- Antrieb"
13. Flight Global, Dominic Perry, "Pipistrel nears launch of new hydrogen-powered 19-seater", 16 February 2021
14. New Atlas, Loz Blain, "Universal Hydrogen and Magnix building world's largest hydrogen plane", September 22, 2020
15. AeroBuzz.De, Volker K. Thomalla, "Airbus arbeitet am emissionsfreien Verkehrsflugzeug" 21.09.2020
16. The Maritime Executive, World's First Liquefied Hydrogen Carrier Launched", Dec 11, 2019
17. J. Töpler, J. Lehmann (eds.), "Hydrogen and Fuel Cells", Springer Verlag Berlin Heidelberg, 2015, Chapter 5, A. Westenberger, "Mobile Application in Aviation" doi 10.1007/978-3-662-44972-1_5
18. M. Janic, "Greening commercial air transportation by using liquid hydrogen (LH2) as a fuel", International Journal of Hydrogen Energy 39 (2014), 16426 - 16441
19. 3rd joint CEP/NOW Heavy Duty Event, April21, 2021 https://www.now-gmbh.de/en/news/events/now-cep-heavy-duty-event/
20. F.J. Edeskuty, "Safety of Liquid Hydrogen in Air Transportation", Deutsche Gesellschaft für Luft- und Raumfahrt/Deutsche Forschungs- und Versuchsanstalt für Luft- und Raumfahrt e.V, International Symposium, "Hydrogen in Air Transportation", Session 4, September 11–14 1979, Stuttgart, Germany
21. Dr. Michael R. Swain University of Miami Coral Cables, FL 33124, "Fuel Leak Simulation", Proceedings of the 2001 DOE Hydrogen Program Review NREL/CP-570-3053
22. A. Marangon et al., "Safety distances: Definition and values", International Journal of Hydrogen Energy 32 (2007) 2192-2197
23. K. Verfondern, "Safety Considerations on Liquid Hydrogen", Schriften des Forschungszentrums Jülich, Energy and Environment, Volume 10, ISBN 978-3-89336-530-2
24. International Journal of Hydrogen Energy, Vol. 19, No. 1, pp. 53–59, 1994, "Large-Scale Hydrogen Liquefaction in Germany", M. Bracha, G. Lorenz, A. Patzelt and M. Wanner
25. International Journal of Hydrogen Energy, Vol. 34, 11560-1568, 2009, "Comparison criteria for large-scale hydrogen liquefaction processes" David O. Berstad, Jacob H. Stang, Petter Neksa
26. International Journal of Hydrogen Energy, XXX, 1–16, 2017, "Process optimization for large-scale hydrogen liquefaction" U. Cardella, L. Decker, J. Sundberg, H. Klein
27. K.Stolzenburg, Prof. em. Dr. H. Quack, "Wasserstoff effizient verflüssigen-Ergebnisse des Projektes IDEALHY", HZwei 01/14
28. T. Numazawa et al., "Magnetic refrigerator for hydrogen liquefaction", Cryogenics 62 (2014) pages 185–192, 2014
29. Viehbahn, Peter (ed.) et al (2018), Technologien für die Energiewende: Technologiebericht—Band 2. Teilbericht 2 zum Teilprojekt A im Rahmen des strategischen BMWi-Leitprojekts

"Trends und Perspektiven der Energieforschung", Wuppertal Report, No. 13.2, Wuppertal Institut für Klima, Umwelt, Energie, Wuppertal
30. Hydrogeit Verlag, www.hzwei.info, Prof. Michael Berger, „Grüner Wasserstoff für die Industrie—Westküste100 nimmt in der Region Heide Arbeit auf", 20. Jahrgang, Heft 4, Oktober 2020
31. Taiki Koide, "Kawasaki Heavy builds world's first tanker for liquid hydrogen", The Asahi Shimbun—Asia & Japan Watch, May 25, 2021
32. G.D. Brewer, editor, "LH2 Airport Requirements Study", NASA contractor report, NASA CR—2700, October 1976
33. Daryl Brown, "Hydrogen supply and Demand—Past, Present and Future", gasworld April 2016, special feature
34. Leigh Collins, "Global green-hydrogen pipeline exceeds 200 GW—here's the 24 largest gigawatt-scale projects", Recharge—Global news and intelligence for the Energy Transition, updated 25 June 2021
35. B. Stalmann, "Wasserstoffzentrum Helgoland", UMWELTMAGAZIN BD.51 (2021) Nr. 07-08
36. Jimmie Langham, managing director AquaVentus, www.aquaventus.org, personal communication July 2021
37. H2 Nationaler Wasserstoffrat, Stellungnahme "Wasserstoff für die Luftfahrt in Deutschland", 16.April 2021 https://www.wasserstoffrat.de/veroeffentlichungen/stellungnahmen-und-positionen
38. https://home.cern/sites/home.web.cern.ch/files/2018-07/CERN-Brochure-2015-003-Eng.pdf. CERN-Brochure-2015-003-Eng March 2015
39. Woodrow Bellamy III, "Universal Hydrogen Eyes Disruptive New Concept to Power Turboprop Aircraft by Mid 2020s", Aviation Today, September 14, 2020
40. Interesting Engineering (https://interestingengineering.com) Ameya Paleja, "Universal Hydrogen makes airlines switch from jet fuel to green hydrogen", July 21, 2021
41. Bruce S. et al. "Opportunities for hydrogen in aviation", CSIRO (Australia's Pre-eminent National Science Organisation) 2020
42. Int. Journal of Hydrogen Energy 26 (2001) 777–782, K. Pehr, P. Sauermann, O. Traeger, M. Bracha, "Liquid hydrogen for motor vehicles—the world's first public LH2 filling station"
43. www.H2-view.com, 11.02.2021, Molly Burgesson, Paris airports could be transformed into hydrogen hubs
44. www.H2-view.com, 05.04. 2021, George Hyneson, Groningen Airport aims to be first hydrogen valley airport
45. www.gasworld.com, 21 June 2021, Molly Burgess, "Air Liquide, Airbus, Groupe ADP prepare airports for the hydrogen era"
46. G. Bernecker, "Planung und Bau verfahrenstechnischer Anlagen", VDI Verlag, Düsseldorf 1984

# Fuel Cells for Unmanned Aerial Vehicles

**Bin Wang and Dan Zhao**

## 1 Introduction

Over the last decades, the global market of unmanned air vehicles (UAVs), also known as drones, has been growing explosively [1, 2]. Especially in China, the average annual growth from 2018 has been even more than 50%. As UAVs have advantages of automaticity, maneuverability, and flexibility, they have been increasingly applied to various civilian, scientific, and military application fields [3, 4]. Generally, small UAVs are widely used to conduct aerial photography in our life. Meanwhile, some small and medium UAVs are also applied in civilian and scientific applications such as crop or forest protection, express transportation, electric power inspection, endangered animal tracking, atmospheric sampling, meteorological observation, and so on. Because of the extremely expensive costs and flight restrictions, large UAVs are usually employed for military applications such as military transportation, dangerous and emergency rescue, fight against terror, etc.

To ensure the long flight duration and specified payload capabilities of UAVs, the selection and design for their power systems are of great importance. In the past, the power systems of UAVs were designed based on internal combustion engine (ICE) drive systems [3, 5]. The ICE drive systems could ensure high-power output for UAVs. Furthermore, the UAV powered by an ICE drive system could achieve a long flight duration since the energy density of the ICE's fuel is very high. However, the

B. Wang
School of Mechanical Engineering, Xi'an Jiaotong University, Xi'an, Shaanxi, China
e-mail: wangbin8751@xjtu.edu.cn

D. Zhao (✉)
Department of Mechanical Engineering, University of Canterbury,
Christchurch, New Zealand
e-mail: dan.zhao@canterbury.ac.nz



C. O. Colpan, A. Kovač (eds.), *Fuel Cell and Hydrogen Technologies in Aviation*,
Sustainable Aviation, https://doi.org/10.1007/978-3-030-99018-3_3

large weight/volume of ICEs would be severe impediments to the miniaturization and lightweight design of UAVs. Moreover, the engine noise and vibration would also degrade the performance of UAVs. In addition, the delayed response of ICE could not meet the automatic control requirements of the autonomous UAVs. As an improvement, hybrid ICE-electric power systems have been proposed for autonomous UAVs [6]. Although the responsiveness of hybrid ICE-electric power systems can be improved, the issues of large weight/volume and engine noise/vibration are not solved, whereas new questions of the complex electromechanical structure and energy conversion consumption from internal energy to electric energy are emerging.

With the rapid development of electric drive and energy storage technologies, the pure electric power systems have become the mainstream power systems for these completely autonomous UAVs. Among the energy storage technologies, Li-ion or Li-polymer batteries are the most widely used power source technologies in UAV applications [2, 3]. Compared with the ICE power systems, the Li-ion or Li-polymer battery power systems have faster response capability, higher operating efficiency with zero-emission, and no noise. Besides, small size and lightweight can be achieved for these UAVs powered by battery power systems. The drawback of the Li-ion or Li-polymer battery power systems is their relatively low energy density. Generally, the flight duration of a battery-powered UAV is only about 20 ~ 40 min. For this reason, the battery power systems are not suitable for these industrial UAVs with long flight duration requirements.

Fuel cells (FCs), as emerging technologies in the areas of energy storage and power supply, have attracted considerable interests from researchers and engineers in recent years [2, 3]. FCs possess the advantages of both ICEs and batteries. The FC power systems can be easily refueled as ICE power systems. Simultaneously, they can directly provide electric power as the battery power systems. For different types of FCs, their working principles, byproducts, and emissions are very different. For hydrogen FCs, they have no carbon emission and the byproduct is only with water. Therefore, the hydrogen FC represents a 100% pure green power technology. For methanol FCs, their byproducts and emissions are only with $CO_2$ and water. Hence, the methanol FCs have no toxic or harmful exhaust emission. However, for solid oxide FCs, the emissions and byproducts might include toxic CO except for $CO_2$ and $H_2O$. Nonetheless, the solid oxide FCs produce less pollution than ICEs. Furthermore, in the solid oxide FC-engine hybrid power systems, the emission of solid oxide FCs can be injected into the burner of the engine again such that the maximum utilization of fuel can be achieved [7].

In addition, the operating temperatures of the aforementioned three types of FCs are also very different. The suggested operating temperature of a hydrogen FC is 40~90 °C. In this temperature range, the operating efficiency of hydrogen FCs is between 40 and 60%. For the methanol FCs, the operating efficiency would be very low if the operating temperature is under 30 °C. With the special materials and workmanship, the operating temperature of methanol FCs could reach more than 120 °C to achieve 40% of operating efficiency [8, 9]. But the costs of the special methanol FC would increase a lot. Different from the hydrogen and methanol FCs, the fuels of the solid oxide FC are diverse. The pure hydrogen, natural gas, and methane can be utilized as the fuel of solid oxide FCs. Usually, the operating temperature of solid oxide FCs would reach 600~1000 °C such that their operating efficiency can reach up to 60%. However, due to the more expensive cost and higher operating temperature compared with the hydrogen and methanol FCs, solid oxide FCs have not become the mainstream choice for civilian UAVs. In effect, the hydrogen and methanol FCs are better choices for these small or medium industrial UAVs compared with solid oxide FCs in civilian applications.

Although the hydrogen and methanol FCs are suitable as power sources for these small or medium industrial UAVs in terms of the costs and operating temperature, the operating efficiency and onboard fuel storage styles should be reconsidered for selecting and designing the practical FC power systems in UAVs. Based on the aforementioned introduction, it can be known that the operating efficiency of the methanol FCs is lower than that of the hydrogen FCs. On the other hand, onboard fuel storage styles in UAVs are noteworthy issues [9, 10]. For methanol FCs, the fuel is liquid and stable so that it can be stored in a conventional bottle or cubic vessel. However, compared with the gas fuel, the liquid fuel leads to lower electrochemical reaction and slower response speed of FC power systems. As a comparison, hydrogen can be stored in a variety of ways. Considerable efforts such as compressed gas, liquefied hydrogen, and chemical storage styles have been devoted to improving the hydrogen storage efficiency per unit volume [2, 3, 11]. Furthermore, the compressed hydrogen and chemical hydrogen generation methods have been demonstrated to be very useful to improve the hydrogen storage efficiency and prolong the UAV's flight duration in practical applications.

A single FC power system in an UAV would face many problems. For instance, FC power systems face long start-up time and slow response issues compared with the battery power systems. What's worse, the discordance between the power consumption and the actual output of FCs would lead to unnecessary energy losses. Therefore, auxiliary/complementary power sources and energy conversion devices should be installed to enhance the operating performance of FC power systems. The auxiliary or complementary power sources for FCs include batteries, ultracapacitors, and solar cells [3, 12]. These FC power systems hybrid with auxiliary or complementary power sources are called FC hybrid power systems. Since two or three power sources are combined in the FC hybrid power systems, suitable and effective hybrid topologies are quite necessary.

For FC hybrid power systems, the topologies and power control strategies are close relatives. The topologies determine the active or passive power control styles. Furthermore, the power control strategies should be implemented based on the energy conversion devices or automatic switches. Usually, the energy conversion devices or units are DC-DC converters with boost, buck, and buck-boost circuits and so on. With the DC-DC converters or automatic switches, the active power control can be achieved. Moreover, the active control objectives are to achieve the maximum energy utilization efficiency, prolong the service-life of FCs, and ensure the system stability [13, 14]. Finally, the power optimization strategies should enhance the comprehensive performance of FC hybrid power systems.

This chapter summarizes current technologies and crucial issues of the commonly used FCs for UAV applications. First, the working principles of three types of FCs are introduced. Then, the onboard hydrogen fuel storage styles are introduced. Furthermore, batteries, ultracapacitors, and solar cells used as auxiliary or complementary power sources are concluded. Meanwhile, topologies and power control of the FC hybrid systems are discussed. Finally, we summarize the deficiency existing and future development directions of the FC technologies.

## 2 Commonly Used Fuel Cells and Their Working Principles

Unlike ICEs and batteries, FCs can generate electric power based on the electrochemical reaction between the fuel and oxygen. Compared with the conventional ICEs, the energy conversion efficiency of FCs is higher. Meanwhile, the energy stored in the fuel (e.g., hydrogen fuel) is higher than the battery stored energy under the same mass condition. So, the time of endurance of a FC power UAV is longer than that of a battery powered UAV. For the reported FCs in UAV applications, they can be classified into three categories based on the types of fuels and electrolyte materials: (1) hydrogen or proton exchange membrane FCs, (2) methanol FCs, and (3) solid oxide FCs. In the following, the working principles and differences of these FCs will be discussed.

### *2.1 Hydrogen FCs*

As its name implies, hydrogen FCs use hydrogen as fuel to produce electric power. From the anode to cathode, the protons would pass through the proton exchange membrane. Meanwhile, the electrons would pass through the power load to the cathode to generate electricity. Based on the operating principle, the hydrogen FC belongs to the type of proton exchange membrane FCs. The electrochemical reactions in a hydrogen FC are shown as follows:

On the anode, $H_2$ changes as

$$H_2 \rightarrow 2H^+ + 2e^- \tag{1}$$

On the cathode, the proton would react with oxygen.

$$\frac{1}{2}O_2 + 2H^+ + 2e^- \rightarrow H_2O \tag{2}$$

The overall reaction of the hydrogen FC is

$$\frac{1}{2}O_2 + H_2 \rightarrow H_2O \quad (3)$$

For the hydrogen FCs, the related technologies such as membrane electrode schemes, proton exchange membrane materials, catalyst formulations, and plate electrodes are relatively mature. Researchers and engineers have enough choices to configure or custom-tailor the catalyst formulations and plate electrodes according to the application conditions of UAVs. Moreover, the biggest advantage of hydrogen FCs is the satisfactory and pragmatic operating temperature range. In addition, the hydrogen FCs are also very beneficial to the lightweight design of FC power systems. However, the hydrogen presents with gas form, the major drawback of hydrogen FCs is the low hydrogen storage efficiency. Another drawback is that the membrane electrodes would degrade quickly, which leads to a short service-life of FCs [15]. Nowadays, many efforts have been devoted to improving the onboard hydrogen energy storage efficiency and the service-life of the membrane electrodes [10, 16]. These efforts would be further discussed in the next section.

## 2.2 Methanol FCs

The methanol FC also belongs to the type of proton exchange membrane FCs. Different from the hydrogen FC, the proton exchange membrane in methanol FCs is the water-rich proton exchange membrane. In the anode, the protons would be produced and then pass through the water-rich proton exchange membrane. Meanwhile, the electrons would also travel through the power load to the cathode to generate electricity.

The electrochemical reaction on the anode is

$$CH_3OH + H_2O \rightarrow CO_2 + 6H^+ + 6e^- \quad (4)$$

Meanwhile, the reaction on the cathode of the methanol FC is

$$\frac{3}{2}O_2 + 6H^+ + 6e^- \rightarrow 3H_2O \quad (5)$$

In summary, the comprehensive reaction of the hydrogen FC is

$$CH_3OH + \frac{3}{2}O_2 \rightarrow CO_2 + 2H_2O \quad (6)$$

Since the methanol fuel presents with liquid form under normal atmospheric temperature conditions, the energy storage efficiency per unit volume is very high. Moreover, the operating temperature of methanol FCs is close to that of the hydrogen FCs, while the methanol production is easier and cheaper than hydrogen production. Meanwhile, the methanol fuel is very stable such that the onboard fuel storage and management can be easily achieved. However, the stable liquid methanol fuel is also a drawback for the electrochemical reaction. Indeed, slow response speed and low operating efficiency cannot be avoided for methanol FCs. What's worse, the catalysts and methanol fuel would lead to poisoning problems. These drawbacks would seriously hamper the popularization and application of methanol FCs. Despite this, methanol FCs have the potential to improve the flight durations for the UAVs with a small fuel storage vessel [9, 17]. Researchers also focus on improving the operating temperature or membrane electrode to achieve the high operating efficiency of methanol FCs [17, 18].

## 2.3 Solid Oxide FCs

Solid oxide FCs belong to the type of high-temperature FCs. The electrolytes of solid oxide FCs are solid oxides such that they can withstand the high operating temperature in the range of 600~1000 °C. In theory, the energy density and power density of solid oxide FCs could be better than that of hydrogen and methanol FCs because solid oxide FCs can use the fuels with high energy/power density to generate electricity. However, due to the characteristics of ceramic materials, solid oxide electrolytes, and high-operating temperature, the miniaturization and lightweight design is very difficult for solid oxide FCs. Nonetheless, researchers have been trying to develop small and low temperature (500–800 °C) solid oxide FCs [19, 20]. In addition, the solid oxide FCs also have their special applications as auxiliary power sources in aerial vehicles or UAVs. For instance, the jet engine integrated with the auxiliary solid oxide FC system can achieve 70~85% of fuel utilization ratio [21]. In the last chapter of this book, the working principles of the solid oxide FC systems and their potential applications will be further introduced in detail.

The comparative analyses of the above three types of FCs are shown in Table 1. It can be known that the hydrogen and methanol FCs have distinct advantages by considering the small size, suitable operating temperature, and environmental friendliness. Although the solid oxide FCs can use various fuels with high operating efficiency, the miniaturized technology is immature in civilian applications. The high operating temperature and large size/weight would be severe drawbacks for UAV applications. Besides, the operating efficiency of the hydrogen FCs is higher when compared with the methanol FCs. Furthermore, the energy storage capacity of hydrogen fuel is lighter than that of the methanol fuel such that the hydrogen FCs are very beneficial for the lightweight design of UAVs. Overall, the hydrogen FC technologies are more mature and practical. Therefore, using the hydrogen FCs to achieve the power supply has become the most popular scheme in UAV applications.

**Table 1** Comparison analyses of the above three types of FCs

| FC types | Advantages | Disadvantages |
|---|---|---|
| Hydrogen FCs | • Small size and no carbon emission<br>• Beneficial to lightweight design<br>• Low operating temperature<br>• Relatively high operating efficiency | • High cost of pure hydrogen<br>• Low volumetric capacity of hydrogen<br>• Rapid degradation of the membrane electrodes |
| Methanol FCs | • Small size and environmentally friendly<br>• Methanol fuel is very stable<br>• Low operating temperature<br>• High energy capacity per unit volume | • Low operating efficiency<br>• Slow response speed<br>• Catalyst or fuel poisoning issues<br>• Extra cost for the fuel supply device |
| Solid oxide FCs | • High operating efficiency<br>• High energy/power density<br>• Various fuels can be used | • High operating temperature<br>• HC or CO emissions<br>• Long-time for preheating and start-up<br>• Miniaturized technology is immature |

# 3 Onboard Hydrogen Fuel Storage Styles

For UAVs, onboard fuel storage and supply styles are noteworthy issues. In addition to the liquid form of methanol fuel, liquefied natural gas, or petroleum-based fuels, most fuels are presented with the gas form. For liquid fuels, the major consideration is the forward flow of water/liquid and adjustable flow rate of liquid fuel. For instance, additional micro-pumps and tubes should be designed for the adjustable flow rate of methanol fuel in FCs. Although the passive methanol supply system can save the additional micro-pumps and tubes, the storage devices of liquid fuels should be installed above the FCs to avoid the reversed flow of liquid fuel. What's worse, the methanol should be mixed with the water, which would tremendously reduce the volumetric capacity of the methanol fuel.

Compared with liquid fuels, hydrogen fuel presents with gas form is better in terms of onboard storage and supply. In practice, hydrogen is the most commonly used fuel in UAV applications. Moreover, hydrogen is not just stored in gas form. Recent progress in onboard hydrogen fuel storage technologies including physical and chemical hydrogen storage styles such as conventional compressed hydrogen gas, cryogenic liquid hydrogen fuel, and hydrogen generation based on chemical hydrides [11, 22]. Among the hydrogen fuel storage and supply styles, the compressed hydrogen gas is the most direct and simplest case for onboard hydrogen fuel storage. Although cryogenic liquid hydrogen is a physical hydrogen storage style, it is not practical compared with the compressed and chemical hydrogen storage styles. In the following, the three hydrogen storage styles would be discussed in detail.

### 3.1 Compressed Hydrogen Style

Usually, three types of hydrogen storage vessels with pressures from 15 to 50 MPa are used for the storage of compressed hydrogen. The I-type hydrogen storage vessel is made of pure metal (steel or aluminum). With the I-type hydrogen storage vessel, the conventional pressure is about 10 MPa and the hydrogen storage efficiency is very low. The II-type hydrogen storage vessel is made of reinforced carbon-around aluminum or steel, the maximum pressure is over 35 MPa such that the hydrogen storage efficiency is improved. To further improve the hydrogen storage efficiency, the III-type hydrogen storage vessel is made of a fully carbon-wound tank with a metal liner, which can withstand the high-pressure over 50 MPa [23]. For the Ion Tiger UAV, the III-type hydrogen storage vessel used in 2009 could achieve 13% of hydrogen storage efficiency (mass ratio of the hydrogen fuel), which is a 10-fold improvement compared with the storage vessel reported in the Spider Lion UAV in 2005 [20].

### 3.2 Cryogenic Liquid Hydrogen Style

In low-temperature conditions, cryogenic liquid hydrogen can be obtained. The cryogenic liquid hydrogen can reach a very high density under the temperature of 20 K, which increases about two times than the compressed hydrogen storage under the pressure of 35 MPa [20, 23]. On the other hand, the pressure of the cryogenic liquid hydrogen fuel is reduced. According to the experimental test of the naval research laboratory, the Ion Tiger UAV using the liquid hydrogen can achieve more than 2 days flight-time [20]. Although the cryogenic liquid hydrogen fuel can achieve high volumetric capacity, the onboard storage of liquid-hydrogen technology is not suitable for widespread use. The thermal insulation for achieving extremely low temperature in the onboard container is very difficult. Meanwhile, the liquid hydrogen evaporation should be carefully controlled, or the high pressure into the liquid hydrogen vessel is very dangerous. For small-scale use, 30~40% of the energy waste for the liquefaction of hydrogen is a big cost. So, the UAV using the liquid hydrogen fuel is only suitable for laboratory research and some special military UAVs.

### 3.3 Chemical Hydrogen Storage Styles

In addition to the cryogenic liquid hydrogen style, the solid chemical hydrogen storage styles can also be used to achieve high volumetric efficiency for hydrogen storage. The major advantage of the solid chemical hydrogen storage styles is that the high hydrogen storage efficiency per unit volume can be achieved at ambient temperature. Meanwhile, the chemical hydrides are very stable and have no hydrogen evaporation issue such that the hydrogen losses can be reduced. The chemical hydrides include metal hydrides, borohydrides, hydrocarbon, and so on. The most widely used chemical hydrides are sodium borohydride and ammonia borane. In the following, the hydrogen extractions based on sodium borohydride and ammonia borane would be discussed. Two methods can be used for the hydrogen extraction from the sodium borohydride ($NaBH_4$) [11]. The simple method is hydrolysis.

$$\mathrm{NaBH_4 + 2H_2O \rightarrow NaBO_2 + 4H_2} \quad (7)$$

The drawback of the hydrolysis reaction is very slow. As a result, the hydrogen extraction rate might not satisfy the demands of the FC power system. To accelerate the hydrogen supply rate, hydrochloric acid (HCl) can be added into the water to react with the chemical hydride $NaBH_4$.

$$\mathrm{NaBH_4} + HCl + \mathrm{3H_2O \rightarrow NaCl + H_3BO_3 + 4H_2} \quad (8)$$

To effectively control the hydrogen supply rate, the water and hydrochloric acid should be regulated and injected into the reactor of $NaBH_4$ [11]. In addition, the volatile portion of hydrochloric acid mixed with hydrogen gas should be eliminated with an acid gas absorbent device. For this reason, an extra pump, water tank, and acid gas absorbent device should be installed. As a result, the mass and size of the hydrogen storage system might be very large and unsuitable for small and medium UAVs.

Compared with sodium borohydride, ammonia borane is a good choice to reduce the mass and size of the hydrogen storage and generation system. The hydrogen generation reaction is without water or the acid solution. Moreover, the hydrogen storage capacity of ammonia borane is higher than that of sodium borohydride. The hydrogen generation reaction of ammonia borane ($NH_3BH_3$) is

$$NH_3BH_3 \rightarrow NH_2BH_2 + \mathrm{H_2} \quad (9)$$

Generally, the heating temperature of 120~180 °C should be provided for the hydrogen generation reaction of ammonia borane. Unfortunately, the hydrogen release from ammonia borane is very slow even in heating conditions. In addition, the foaming phenomenon would lead to the plugging issue into the ammonia borane reactor. Despite these difficulties, the feasibility of the UAV with the hydrogen storage and supply system based on ammonia borane was demonstrated in 2013 [10]. Results showed that the flight duration of this UAV could reach about 1 h.

Although the chemical hydrogen storage styles have drawbacks such as the byproduct, long-time heating, large onboard hydrogen fuel storage vessel, etc., scientists have tried to look for high-efficiency chemical hydrides and catalysts to achieve fast and reliable onboard hydrogen generation. The solid chemical hydrogen storage styles are also very popular in the applications of military UAVs.

# 4 Auxiliary or Complementary Power Sources

According to the comparisons of the operating characteristic, technology maturity, and practicability among the hydrogen FCs, methanol FCs, and solid oxide FCs, adopting the hydrogen FCs as major power sources would be the best scheme in UAV applications. Nowadays, the improved catalyst formulation, the composite bipolar plate, and the automatic humidifier make the hydrogen FCs adapt to various application scenarios [8, 24, 25]. Despite the amazing advances in hydrogen FC technologies, pure hydrogen FC power systems without auxiliary or complementary power sources have their drawbacks to provide propulsion power for UAVs. The pure hydrogen FC power systems cannot meet the fast-response and high-power requirements of the flight load, especially in the case of emergency takeoff and crash landing. If hydrogen FCs are forced with high variation in power demand, the strong electrochemical reactions would occur into FCs, which would permanently reduce their operating efficiency and even shorten their service-life. In this case, the auxiliary or complementary power sources are suggested to enhance the response capability, improve the operating efficiency, and prolong the service-life of hydrogen FC power systems.

## *4.1 Batteries as Auxiliary Powers*

Nowadays, batteries have become the most widely used power source in various vehicles due to their high-power density and relatively high energy density [2, 26]. Indeed, a pure battery power system can only achieve 20~30 min of flight endurance for industrial UAVs. Although some specific batteries have higher energy density such that the UAV power by them have long flight endurance, the power density of these specific batteries would be reduced obviously. As a result, the load-carrying ability of the UAV is low and unsatisfactory. Due to the limitations of the battery materials and manufacturing process, a battery cannot possess both high power density and high energy density. In addition, most batteries have poor performance and capacity fade issues in low-temperature conditions.

The specific energy/power comparisons among different power supply technologies are illustrated in Fig. 1. We can see that the energy density of batteries is much lower than that of the FCs. However, batteries have about 5~10 times the power density of FCs. Therefore, in the FC/battery hybrid power system, the FCs can be used as the major power source to meet the continuous power supply requirement. Meanwhile, batteries can be utilized as an auxiliary power supply to meet the fluctuating power demand. Moreover, batteries can also ensure a stable voltage output for the FC/battery hybrid power system.

In the last 10 years, the Li-ion or Li-polymer batteries as the auxiliary power source of FC power systems in UAVs have been widely studied [27–29]. For instance, a small UAV equipped with a special FC/battery hybrid power system was

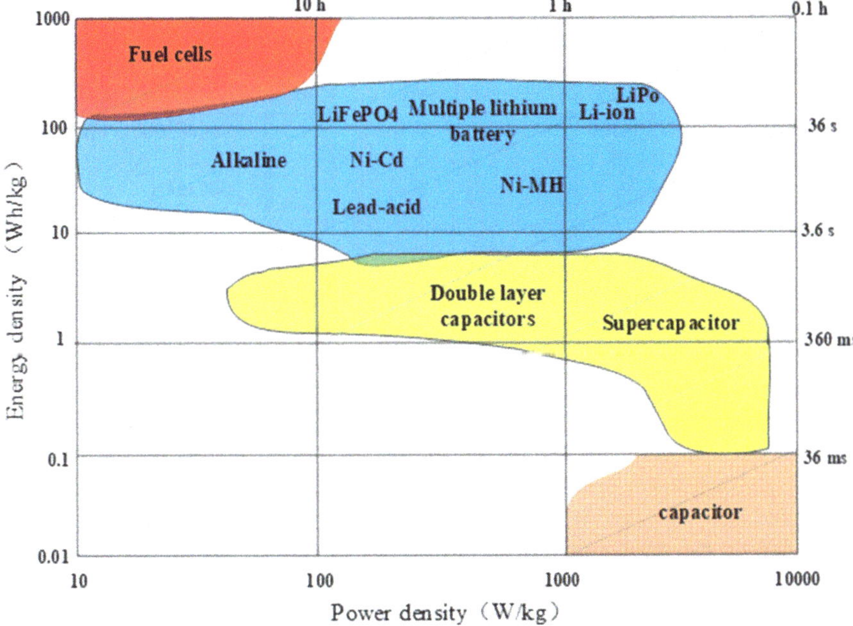


**Fig. 1** Specific energy/power comparisons among different power supply technologies

studied in 2013 [30]. In that study, the FC/battery hybrid power systems conducted the hybrid output mode to meet the large power requirement. If the power requirement is low, the pure FC output mode was conducted. Usually, the energy losses of the DC-DC conversion can be avoided by using the pure FC output mode during the cruise flight mode of the UAV. During the climb or descent modes, the battery can adjust the output power of the FCs and ensure the large-power output ability of the overall FC power systems. In the literatures [29, 31], further evidence showed that the battery could adjust the FC output power to achieve the maximum hydrogen energy conversion efficiency in UAV applications.

## *4.2 Supercapacitors as Auxiliary Power Sources*

Supercapacitor, characterized by very high power density as a power supply device, has attracted quite a lot of attention over the last 10 years [3, 13]. It can be known from Fig. 1 that the power density of the advanced supercapacitor is improved over twice that of the best-performance batteries (i.e., Li-ion and Li-polymer batteries). The performance comparisons including the specific energy/power, the cycle-life, and the operating efficiency among the advanced supercapacitor and two best-performance batteries are illustrated in Table 2. Apart from the high-power density, it can be known that the supercapacitor has an incomparable cycle-life, which is

**Table 2** Comparisons among the advanced supercapacitor and two best-performance batteries

| Power sources | Energy density (Wh/kg) | Power density (W/kg) | Life cycle | Efficiency |
|---|---|---|---|---|
| Li-polymer battery | 150~200 | 2500 | Up to 1200 | Up to 90% |
| Li-ion battery | 150~300 | 2000 | Up to 1000 | Up to 92% |
| Supercapacitor | 10~20 | 5000~15,000 | Up to 100,000 | Up to 98% |

very suitable for high-frequency charging and discharging applications. What's more, the supercapacitor can suffer from the low-temperature environment. Because of this advantage, using a supercapacitor pack in the FC hybrid power system can address the poor power performance and capacity fade issues of power sources in extremely cold applications.

Indeed, if UAVs or unmanned aircrafts fly at a high altitude, the extremely cold weather would directly degrade the power performance of FCs and batteries. What's worse, the high-power operating in low-temperature conditions would damage batteries and FCs such that the service-life of the overall FC hybrid power system is shortened. In this case, adding a supercapacitor module to supply auxiliary power can effectively enhance the power performance and prolong the service-life of the FC hybrid power system, simultaneously.

Unlike batteries, the supercapacitor has a specific structure with double-layer, which results in fast response speed [32, 33]. Therefore, supercapacitors are the ideal power supply devices to deal with the sudden load changes during emergency landing or climbing conditions in UAV or aircraft applications. Meanwhile, the supercapacitor resistance is extremely small. The operating efficiency of supercapacitor can reach up to 98%. Although the supercapacitor has so many advantages, its specific energy is less than 10% of that of the battery. For this reason, the size and weight of an auxiliary power source with the pure supercapacitor system might be very unsatisfactory. Usually, the supercapacitor should be integrated with the battery as a hybrid energy storage system to provide the auxiliary power output for FCs. For instance, in the FC/battery/supercapacitor hybrid power system of more-electric aircrafts, the supercapacitor directly connects the DC-bus to filter the high-frequency and large-pulse power, while the battery is a power and energy buffer used to ensure the stable power output for the DC-bus [33, 34]. With this design, stable and optimal power output can be realized for this FC hybrid power system.

Using a supercapacitor as an auxiliary power source in UAV applications also has some drawbacks. The supercapacitor output voltage is not stable as that of the battery. The direct connection between the supercapacitor and DC-bus would go against the stable control of the FC hybrid power system. For this reason, a voltage regulator is suggested to be installed for adjusting the voltage of the DC-bus. However, the voltage regulator would further increase the size and weight of the power system. In addition, the self-discharge effect of the supercapacitor would increase the energy losses. Therefore, in UAV applications, especially for small UAVs, the high-power batteries are suggested to replace the supercapacitor to supply auxiliary power. For these UAVs in low-temperature applications, the supercapacitors are suggested to be integrated with the batteries to provide the auxiliary power, while the heating measures should be taken for batteries. Fortunately, the waste heat produced by the FCs can be utilized for heating the overall FC/battery/supercapacitor hybrid power system.

## 4.3 Solar Cells as Complementary Power Sources

Batteries and supercapacitors need to be charged in advance, and their stored energy is limited. Accordingly, their discharge time is also limited. In terms of long endurance, the solar cells would be a great choice as a complementary power source for FCs in UAV applications. In contrast to the batteries and supercapacitors, solar cells do not need to be charged in advance and the solar energy capture from the sun is unlimited. For high-altitude long-endurance (HALE) UAVs, enough sunlight can be obtained since there is no cloud cover when the HALE UAVs fly high enough [2, 35]. On the other hand, For the UAV carried an FC/solar cell/battery hybrid power system, the extra solar energy can be stored in batteries for supporting the night flight. The power/voltage output performance of solar cells is also unstable, i.e., the solar cell power system has low power performance and decaying voltage phenomena along with the increase of operating current. Therefore, the solar cells cannot be utilized as an auxiliary power source to supply instantaneous power or stable voltage. On the contrary, the solar cells should be in parallel with the FCs to supply complementary power.

The solar cells should be placed upon the two wings of UAVs. In this aspect, the FC/solar cell/battery hybrid power systems are not suitable for pure rotary-wing UAVs. Meanwhile, a maximum Power Point Tracking (MPPT) controller integrated with batteries and FCs should be installed in the middle of the fuselage to ensure the balance gravity center of UAVs [3, 14]. Notice that, the output power from the solar cells would charge the batteries. An energy management strategy is required to ensure the charge and discharge safety of the batteries. For FC/solar cell/battery hybrid power systems, their energy management strategy can be implemented according to the MPPT controller. For instance, when the battery voltage is equal to its maximum voltage, the MPPT controller should be operating with the constant voltage output mode.

In a previous report, some researchers studied the passive connection styles of the FCs, batteries, and solar cells [36]. With the passive connection styles, the difference among the output voltages of the three power sources would be a big problem. Although the reversed-biased power diode can solve this problem, the clamping voltage might lead to invalid output for the FCs or solar cells. Nonetheless, the previous studies demonstrated that the solar cells in the passive FC/solar cell/battery hybrid power system can improve the flight duration of UAVs [36]. In recent years, many improved studies have tried to enhance the operating performance or efficiency of the FC/solar cell/battery hybrid power system [3, 14]. Nowadays, the

UAVs with the FC/solar cell hybrid power system can conduct more than 1-day flight endurance and even realize the global flight task.

In summary, FCs can be integrated with other power sources as hybrid power systems to supply power for UAVs. Due to the different operating characteristics of batteries, supercapacitors, and solar cells, they play auxiliary or complementary roles for FCs. FCs could be used as major power sources, while the battery is the vital auxiliary power supply device. To enhance the high-power output performance and the low-temperature performance, a supercapacitor can be added as the auxiliary power source. For large UAV applications without the weight limitation, supercapacitors are recommended as a power buffer to absorb/supply the high-frequency and large-pulse power. In this case, the power fluctuation of the battery and FC can be restrained. In addition, solar cells with low power performance cannot be considered as an auxiliary power source. Since the energy produced by the solar cells is unlimited, solar cells are suggested to be used as a complementary or parallel power supply of FCs to achieve long-endurance for UAVs.

## 5 Hybrid Topologies and Power Control

Hybrid topologies determine the power control and output performance of the FC hybrid power systems. The hybrid topology design needs to consider the following factors: (1) primary, complementary, and auxiliary relations among different power sources, (2) active or passive power control for power sources, and (3) protection and service-life extension for the primary power sources. The primary and auxiliary relations are related to the parameter/size matching and optimization of different power sources. Furthermore, DC-DC converters or automatic switches can be used to actively control the primary or complementary power sources, while the batteries and supercapacitors used as auxiliary energy or power pumps can be passively controlled. In addition, to protect the primary power source and prolong its service-life, excessive power output should be suppressed or avoided by actively controlling the DC-DC converters or automatic switches.

### *5.1 Passive Topologies*

Passive topologies are the simplest integrated patterns to achieve the hybrid power output for FCs and their auxiliary/complementary power sources. The advantage of the passive topologies is that the DC-DC conversion losses can be completely avoided. However, the voltage coupling or incompatibility among different power sources is a noteworthy issue to passive FC hybrid power systems. Besides, the reverse power which would shock FCs and solar cells must be avoided to realize the service-life extension of the overall hybrid power system.

(a)

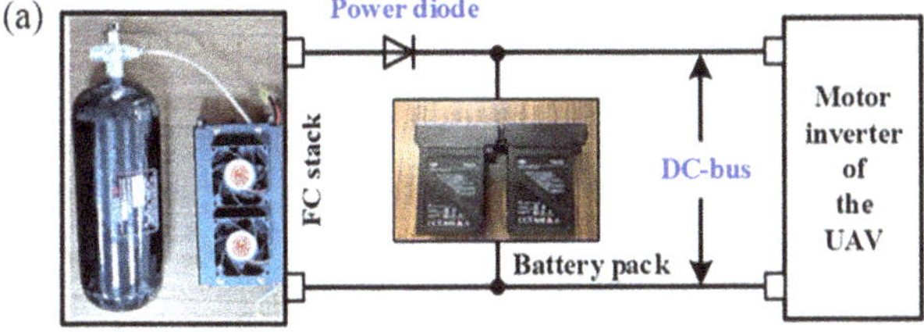


(b)

**Fig. 2** Two passive FC hybrid power systems. (**a**) Passive topology of the FC/battery hybrid power system. (**b**) Passive topology of the FC/solar cell/battery hybrid power system

Two passive FC hybrid power systems are shown in Fig. 2. It can be seen that the power diodes are installed in the power-flow paths of the FC stack or the solar cells to avoid the reverse power to shock the FCs. With the passive topologies, the output power/voltage of the FC stack or the solar cells cannot be actively controlled and automatically adjusted. The FCs or solar cells would generate power only if their output voltage is higher than that of the battery voltage.

To assure that the FC can operate with the optimal output voltage, the maximum battery voltage needs to be close to the optimal output voltage of the FC stack for the passive FC/battery hybrid power system. However, the utilization efficiency of the battery stored energy might be very low. For the FC/battery/solar cell hybrid power systems, the solar cells cannot maintain a usable voltage range when the weather is not very good. In this case, the utilization efficiency of solar energy is also very low. In the following, the passive topologies would be improved with DC-DC converters or automatic switches to realize a high utilization ratio of the battery stored energy and solar energy.

## 5.2 *Semi-Active Topologies*

To improve the utilization efficiency of the battery stored energy and the solar energy, engineers designed the semi-active topologies for the FC hybrid power systems. In these semi-active topologies, one of the power sources should be actively controlled. In general, the DC-DC converters are used to achieve the output voltage adjustment and power control for the FCs or solar cells. As illustrated in Fig. 3a, a DC-DC converter is installed between the FC stack and the battery pack. In general, the DC-DC converter is the boost converter that can achieve a high output voltage. Meanwhile, the input voltage and current can also be actively adjusted by the boost converter such that a stable and optimal output power can be achieved for the FC stack. In this case, both the power generation efficiency and the service-life of the FC stack can be increased.

For UAV applications, an alternative semi-active FC hybrid power system is the semi-active Battery/FC hybrid power system. This alternative semi-active topology exchanges the passive and active control roles for the FC stack and the battery pack, as shown in Fig. 3b. Notice that the battery pack can be charged and discharged,

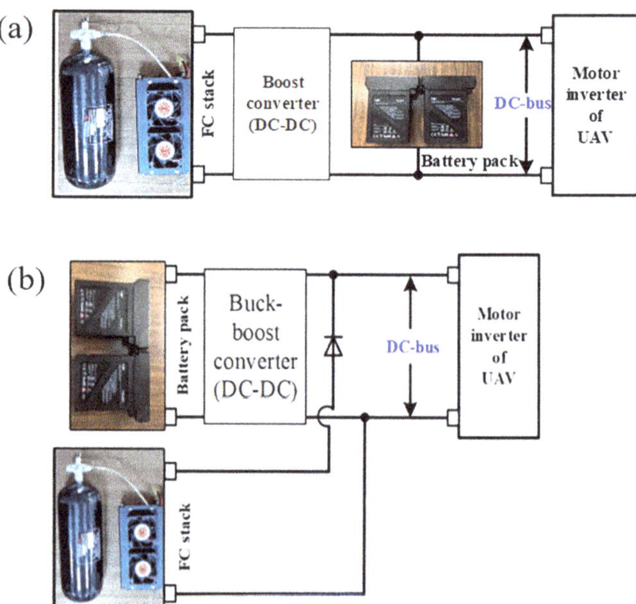


**Fig. 3** Semi-active FC hybrid power systems. (**a**) Semi-active FC/battery hybrid power system. (**b**) Semi-active Battery/FC hybrid power system

simultaneously. Therefore, a bidirectional DC-DC converter should be installed for adjusting the input/output power of the battery. The bidirectional DC-DC converter is a buck-boost converter that can also ensure a stable input voltage for the motor inverter. In addition, the battery pack can be charged with this buck-boost converter when the power demand from the motor inverter is very low. With this topology, the FC stack can supply power to the motor inverter without the energy losses in the DC-DC converter. However, an obvious drawback is that the FC stack might be damaged by the instantaneous pulse power. In addition, the battery pack would be charged and discharged frequently, which would also increase the energy losses in the DC-DC converter.

Solar cells as a complementary power source would play a very different role in FC hybrid power systems compared with the battery pack. The solar cells have no such function of energy storage as the battery pack. On the other hand, the output voltage performance of the solar cells is similar to the FCs. Hence, using the battery pack as the auxiliary power source in the FC/solar cell hybrid power system is very necessary. Moreover, to achieve the maximum power generation efficiency, a maximum power tracking (MPPT) converter should be employed for adaptively adjusting the power output of solar cells. The semi-active FC/solar cell/battery hybrid power system is illustrated in Fig. 4.

Sometimes, the supercapacitor can be integrated with the FC stack and battery pack to enhance the power output characteristic of the FC hybrid power system [3, 33]. Especially, the low-temperature operating characteristic of the supercapacitor could improve the low-temperature performance of the overall FC hybrid power system. However, supercapacitor voltage is unstable such that the power control of the supercapacitor is extremely unsteady. To solve this problem, the DC-DC converter is essential to ensure the stable power output for the supercapacitor. Although the supercapacitor can enhance the power output characteristic of the FC hybrid power system, the DC-bus voltage might not be stable with only the supercapacitor. Therefore, the FC hybrid power system still needs the battery pack to ensure the voltage stability of the DC-bus. Similar to the semi-active FC/solar cell/battery

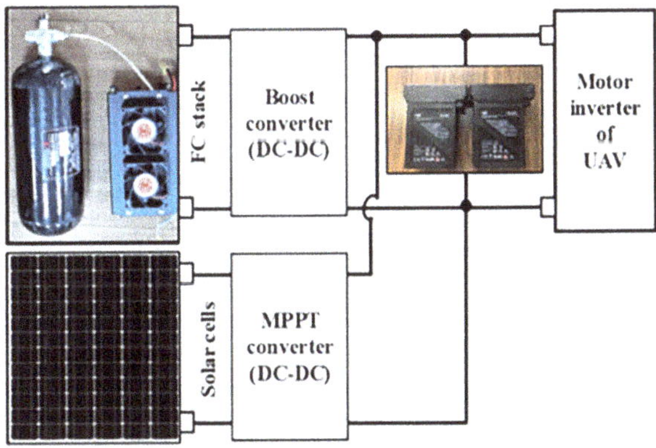


**Fig. 4** Semi-active FC/solar cell/battery hybrid power system

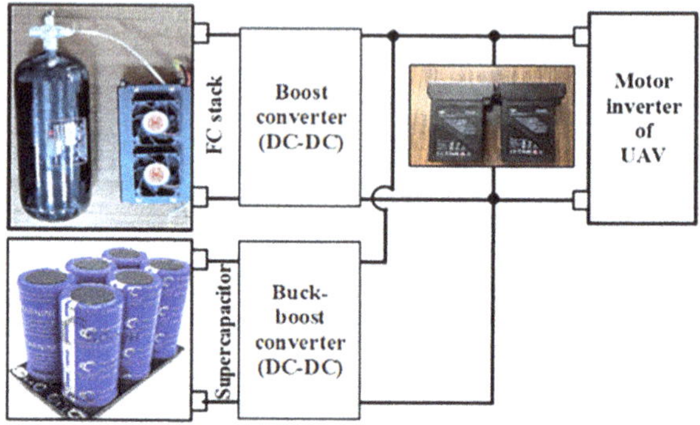


**Fig. 5** Semi-active FC/battery/supercapacitor hybrid power system

hybrid power system, the semi-active topology of the FC/battery/supercapacitor hybrid power system is shown in Fig. 5.

## 5.3 *Fully Active Topologies*

To realize that both the FC stack and the battery pack can be actively controlled, double DC-DC converters were proposed to realize the full control of the FCs and batteries, respectively [3]. As shown in Fig. 6, the two power sources can be fully controlled by their respective DC-DC converters. So, this FC hybrid power system is named the fully active FC/battery hybrid power system. In this topology, there is no interference of the output voltage or power between the FC stack and the battery pack. However, using double DC-DC converters would lead to overweight and large volume. Ultimately, the load capability and flight endurance of the UAV would be reduced. The same problems would also arise for other fully active FC hybrid power systems.

## 5.4 *Application Examples*

By considering the factors of size, weight, cost of the aforementioned topologies, the passive and semi-active topologies are more practical than the fully active topologies for the FC hybrid power systems in small and middle UAV applications. For instance, the weight of a small UAV is within 15 kg, a fully active topology would tremendously affect the mass ratio of FC/battery hybrid power system. What's worse, the loading capacity of this small UAV would be significantly decreased. This section will present two examples about the passive and semi-active topologies for the FC/battery hybrid power systems in fixed-wing and rotary-wing UAVs, respectively.

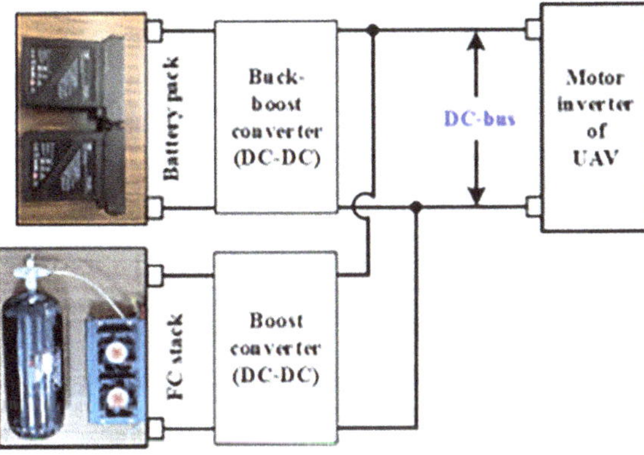


**Fig. 6** Fully active FC/battery hybrid power system

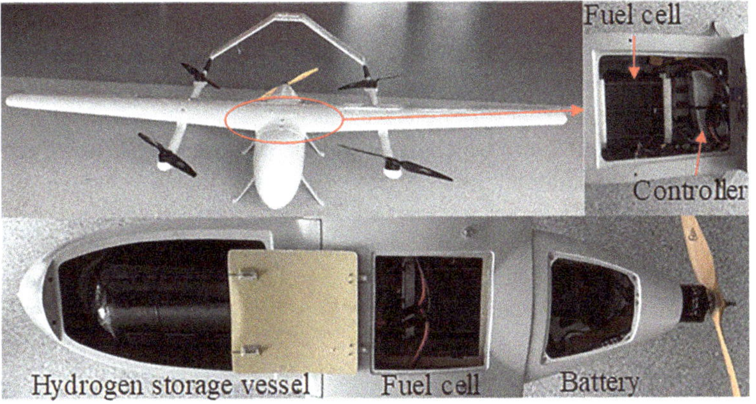


**Fig. 7** Fixed-wing UAV with the FC/battery hybrid power system

For the fixed-wing UAV in Fig. 7, the power requirement after take-off is very low. The low power operation has a less reverse effect on the service-life of the FC. Therefore, it is expected that the FC can directly supply energy to the motor inverter since the DC-DC conversion from the FC to the battery would increase extra energy losses. In this respect, the passive topology is a suitable choice for the FC/battery hybrid power system in the fixed-wing UAV. However, it is known that the passive FC/battery hybrid power system cannot achieve the active power control of the battery pack and the FC stack. To avoid that the battery pack is over charged, the maximum voltage of the FC stack should be equal to the maximum voltage of the battery pack. In addition, an automatic switch is designed in the power-flow path of the FC stack.

Indeed, the automatic switch is a MOSFET, as illustrated in Fig. 10a. The ON/OFF state of this MOSFET is determined by the power requirement from the motor inverter. Because the battery can assure a relatively stable voltage, we can use current sampling to replace the power sampling. Hence, we design a referenced current to achieve the power control. If the sampling current is higher than the referenced current, the power requirement would be very high. This situation occurs during the take-off, emergency stop, or sudden direction change. In this situation, the MOSFET is with OFF state and the pure battery mode is executed. When the fixed-wing UAV is in normal flight, the MOSFET is with ON state and the passive FC/battery mode is executed. With the passive FC/battery mode, if the battery is fully charged, the FC would separately provide power to the motor inverter. Sometimes, with the high-frequency change of the power requirement, the hybrid power output would be passively implemented. With the passive FC/Battery mode, the charge and discharge states of the battery pack are uncontrollable. The two operating modes and power flow control of the passive FC/battery hybrid power system are illustrated in Fig. 8b, c.

For the rotary-wing UAV illustrated in Fig. 9, the power requirement from the motor inverter after take-off is relatively high. The improved passive FC/battery hybrid power system cannot effectively prevent the FC from the high-frequency and pulse power. Hence, the semi-active FC/battery hybrid power system is recommended for this rotary-wing UAV. The FC stack is located at the central and bottom position of the hybrid power system. The electronic devices of the power system are installed in the middle layer, which can avoid the corrosion of the water generated by the FC stack. In addition, the hydrogen vessel is on the uppermost layer of the hybrid power system, which can be replaced easily.

In this semi-active FC/battery hybrid power system, the maximum voltage of the FC stack is less than the rated voltage of the battery pack. The boost converter is

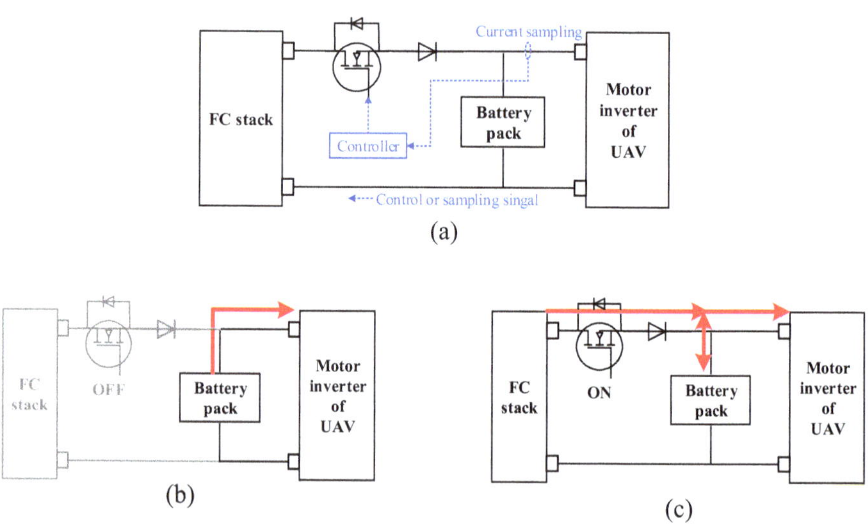


**Fig. 8** Improved passive FC/battery hybrid power system and its power control. (**a**) Improved passive topology of the FC/battery hybrid power system. (**b**) Pure battery mode. (**c**) Passive FC/battery mode

**Fig. 9** Rotary-wing UAV with the FC/battery hybrid power system

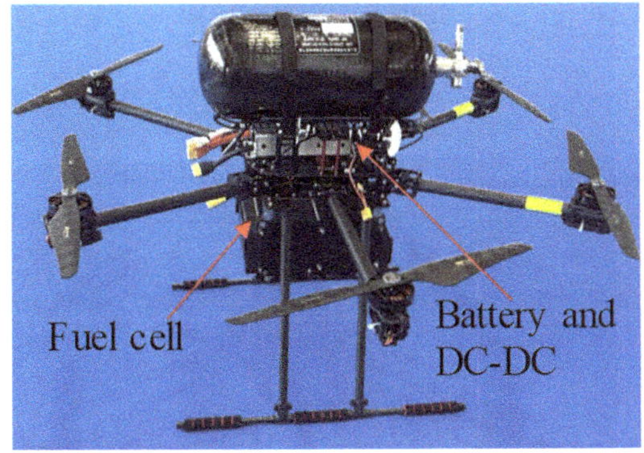


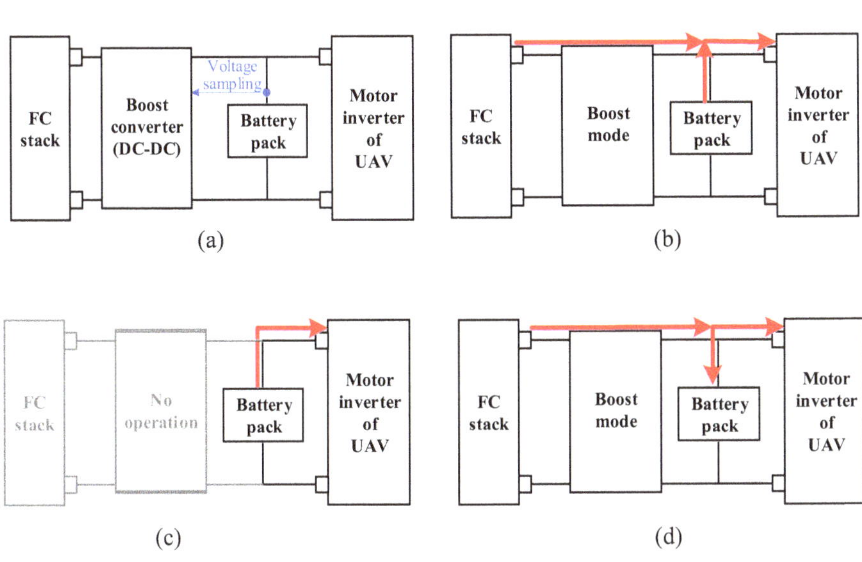


**Fig. 10** Semi-active FC/battery hybrid power system and its operating modes. (**a**) Semi-active FC/battery hybrid power system. (**b**) Hybrid output mode. (**c**) Pure battery mode. (**d**) Battery voltage compensation mode

used to achieve the energy conversion from the FC stack to the battery pack. To simplify the control strategy, the boost converter would regulate the power output of the FC stack according to voltage sampling of the battery, as shown in Fig. 10a. When the battery voltage is in a suitable range, the FC stack can work with the optimal power with the hybrid output mode. If the voltage of the battery pack is too high, the pure battery mode should be executed to avoid overcharging the battery pack. In addition, if the battery voltage is too low, the battery voltage compensation mode should be implemented. With this operating mode, the boost converter should increase the output power of the FC to charge the battery. Obviously, with this semi-active topology, the charge and discharge states of the battery can be actively

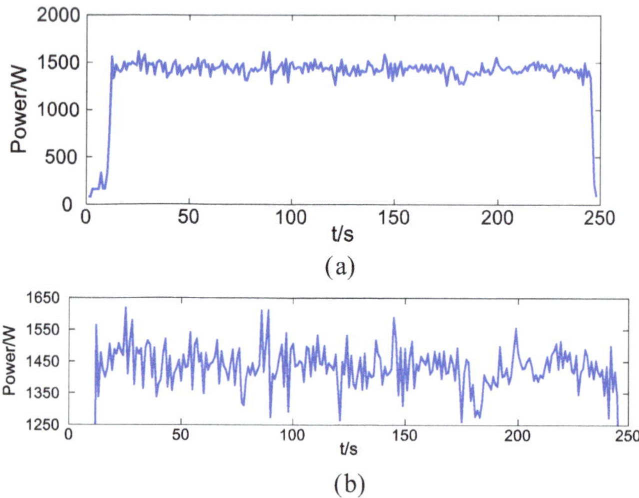


**Fig. 11** Power profile of the rotary-wing UAV

controlled. The three operating modes of the semi-active FC/battery hybrid power system are shown in Fig. 10b–d.

To show the importance of the active control for the FC/battery hybrid power system, the power requirement of the rotary-wing UAV is presented, as shown in Fig. 11a. The power level of our rotary-wing UAV is 2 kW. The optimal power region of the FC in this rotary-wing UAV is 1350~1550 W. Accordingly, the rated voltage of the FC is 48 V, while the operating current can reach up to 60 A. However, it can be seen from Fig. 11b that the power fluctuation after the start-up of the UAV is in the region of 1260~1620 W. Notice that, this power profile has no special operation conditions such as start-up, emergency stop, or sudden direction change with full load, or the power fluctuation would be larger. In this case, if the FC is passively controlled, the FC would not operate with the optimal power. What's worse, the large power fluctuation would directly shock the FC, which might damage the FC seriously. Therefore, for this rotary-wing UAV, the semi-active FC/battery hybrid power system would be better to balance the power between the FC and the battery than the passive FC/battery hybrid power system.

# 6 Crucial Issues for Current Fuel Cell Technologies

## *6.1 High Cost of Fuel Cell–Powered Systems*

Although various FC-powered UAVs have been studied and reported since 2003, the FC technologies have not been in widespread use for commercial UAVs. The first critical issue for impeding the industrialization of FC-powered systems in UAV applications is the high cost. At first, the noble metal catalysts used for electrochemical reactions lead to the high manufacturing cost of FCs. Meanwhile, the membrane electrodes of FCs can be easily degraded or even severely damaged in high-frequency and large-power operation conditions. Although an intelligent controller can suppress the high-frequency and peak-power output for FCs, it increases the complexity and cost of the overall FC-power system. The maintenance costs of FCs are also extremely high. In addition, a pure FC-powered system cannot meet the complex power demand of UAVs. The FC-powered system should be integrated with the DC-DC converter and batteries. The extra DC-DC converter and batteries would also increase the total cost of FC-powered systems.

For commercial UAVs, the Li-ion or Li-polymer batteries are usually used as power sources. As a comparison, the costs of the power system based on Li-ion or Li-polymer batteries would be less than 1000 dollars, while the FC-powered system (i.e., the complete system including the stack, the hydrogen tank, the hybrid controller, and the battery module) would be higher than 16,000 dollars. The general consumers cannot accept this cost gap between the two power systems. Although the FC-powered system can achieve a longer duration than the battery-powered system, using two or three groups of battery-powered systems in turn to supply power is more popular and broadly acceptable. Therefore, for the industrialization of FC-powered systems in UAV applications, engineers need to look for cheaper membrane materials and catalysts to reduce the manufacturing and maintenance costs of FCs.

## *6.2 Low Performance under Cold Environment*

FCs need a suitable operating temperature range to ensure effective electrochemical reactions. Under an extremely cold environment, the power generation efficiency of the FC would decrease severely. Especially, the FC cannot be started in normal mode. Thus, extra preheating devices should be designed for the start-up of the FC. What's worse, if the ambient temperature is less than −30 °C, the power performance of FCs would be reduced by over 10%. When the power performance is reduced, the load-carrying ability of UAVs would be simultaneously reduced. This is also a crucial issue for UAV applications. For example, in the northeast of China, the ambient temperature is usually less than −30 °C in winter, and the power degradation would lead to the low load-carrying ability for UAVs. In our lab tests, if the ambient temperature is less than −40 °C, the UAVs powered by FCs would be unable to take off.

For the FCs, the enhanced metal bipolar plate can improve the low-temperature performance [25]. The drawback is that the metal plate can be corroded by the water and acid-containing liquid such that the service-life is shortened. On the other hand, for the FC/battery hybrid power system, the battery pack plays the role of preheating the FC stack in the start-up phase. However, the battery also faces the power degradation issue in a cold environment. Although some all-climate batteries have been proposed to address the degradation issue in a cold environment, the self-preheating would consume extra energy. As a result, the flight duration of UAVs would be reduced. To address the low performance issue of the FC-power systems under the cold environment, scientists are focusing on looking for the new electrode materials of FCs and batteries. The new electrode materials should have high electrochemical activity such that the conductivity of FCs and batteries can be improved under low temperature. In addition, scientists are also looking for new catalyst formulation and bipolar plate technologies to improve the performance of FCs under the cold environment.

### *6.3 Rapid Performance Degradation in Practical Environment*

Rapid performance degradation is also a big issue for the industrialization of FC-powered systems. To ensure the practicability of the FCs, engineers expect that the FCs can achieve over 10,000 h of service-life in practical applications. For industrial UAVs, the special practical environment would be much worse than the experimental environment. For instance, the FCs in the UAVs used for plant protection would work with toxic insecticide spray, dense smoke, and so on. Because the FCs need oxygen to take part in the electrochemical reaction, the membrane electrodes are directly exposed to the air. Some acid, alkaline, or hybrid polluted gas would be directly absorbed by the membrane electrodes of the FCs. Accordingly, the catalysis of the electrochemical reaction is reduced. What's worse, the catalytic activity of the polluted catalysts would be permanently reduced. In other words, the membrane electrodes would face irreversible performance degradation. Finally, the rapid performance degradation leads to a short service-life for the FC-powered systems.

Usually, the service-life of the membrane electrodes in FCs is about 5000~6000 h in general service environment. In the acid or seriously polluted environment, the service-life of the membrane electrodes is reduced obviously. If the membrane electrodes are seriously damaged, the FCs should be sent back to the manufacturing factory to re-change the membrane electrodes. The consumer would be bored by the long-term maintenance issues.

## 7 Conclusions and Breakthrough Directions in Future

Using FCs as major power sources to achieve long-endurance has become a popular research direction in the field of UAV applications. According to different application requirements, FCs with different working principles as major or auxiliary power sources were proposed and validated in UAV applications in previous studies. In terms of current energy storage and power supply technologies, the FCs have higher energy density because they can use onboard fuel storage systems, while their power density and response capability are not so good compared with batteries and supercapacitors. Therefore, FCs should be integrated with other power sources to obtain satisfactory operating performance for various UAV applications. Moreover, FC hybrid power systems should match suitable topologies according to different types of UAV applications. Although different FC hybrid power systems have been studied extensively in the field of aerial vehicle applications, there are still some crucial issues that need to be urgently addressed for the industrialization of FC-powered UAVs.

This chapter reviewed three reported FCs in vehicle applications at first. Among them, the hydrogen FC is the most suitable choice as the power source for UAVs in terms of clean and environmental protection, small size and lightweight, suitable operating temperature, diversified onboard hydrogen fuel storage technologies, and so on. Then, we discuss the auxiliary or complementary power sources for FCs. In FC hybrid power systems, the battery pack is an essential auxiliary power source to guarantee the stable and reliable voltage/power output of the overall system. On the other hand, the supercapacitors can provide/absorb instantaneous pulse power and the solar cells can provide complementary power output for the FCs, respectively. Furthermore, the feasible topologies for FC hybrid power systems are analyzed. We conclude that the passive or semi-active FC hybrid power systems are more suitable for UAV applications compared with the fully active FC hybrid power systems. Finally, we propose some crucial issues of the promotion and application for current FC technologies in UAV applications. To address these crucial issues, new and low-cost catalyst formulations, adaptive preheating or all-climate FCs based on improved metal plate electrodes, and long service-life based on the novel manufacturing process of membrane electrodes will be the breakthrough directions of FC technologies in the future.

## References

1. A. S. Saeed, A. B. Younes, C. Cai, and G. Cai, "A survey of hybrid Unmanned Aerial Vehicles," Progress in Aerospace Sciences, vol. 98, pp. 91–105, 2018. 04/01/2018.
2. M. N. Boukoberine, Z. Zhou, and M. Benbouzid, "A critical review on unmanned aerial vehicles power supply and energy management: Solutions, strategies, and prospects," Applied Energy, vol. 255, p. 113823, 2019. 12/01/2019.
3. B. Wang et al., "Current technologies and challenges of applying fuel cell hybrid propulsion systems in unmanned aerial vehicles," Progress in Aerospace Sciences, vol. 116, p. 100620, 2020. 07/01/2020.
4. C. De Wagter et al., "The NederDrone: A hybrid lift, hybrid energy hydrogen UAV," International Journal of Hydrogen Energy, vol. 46, no. 29, pp. 16003–16018, 2021. 04/26/2021.
5. Y. Xie, A. Savvarisal, A. Tsourdos, D. Zhang, and J. Gu, "Review of hybrid electric powered aircraft, its conceptual design and energy management methodologies," Chinese Journal of Aeronautics, vol. 34, no. 4, pp. 432–450, 2021. 04/01/2021.
6. B. J. Brelje and J. R. R. A. Martins, "Electric, hybrid, and turboelectric fixed-wing aircraft: A review of concepts, models, and design approaches," Progress in Aerospace Sciences, vol. 104, pp. 1–19, 2019. 01/01/2019.

7. Z. Ji, J. Qin, K. Cheng, H. Liu, S. Zhang, and P. Dong, "Thermodynamic analysis of a solid oxide fuel cell jet hybrid engine for long-endurance unmanned air vehicles," Energy Conversion and Management, vol. 183, pp. 50–64, 2019. 03/01/2019.
8. C. Y. Wong et al., "Additives in proton exchange membranes for low- and high-temperature fuel cell applications: A review," International Journal of Hydrogen Energy, vol. 44, no. 12, pp. 6116–6135, 2019. 03/01/2019.
9. Ó. González-Espasandín, T. J. Leo, M. A. Raso, and E. Navarro, "Direct methanol fuel cell (DMFC) and H2 proton exchange membrane fuel (PEMFC/H2) cell performance under atmospheric flight conditions of Unmanned Aerial Vehicles," Renewable Energy, vol. 130, pp. 762–773, 2019. 01/01/2019.
10. J.-E. Seo et al., "Portable ammonia-borane-based H2 power-pack for unmanned aerial vehicles," Journal of Power Sources, vol. 254, pp. 329–337, 2014. 05/15/2014.
11. S.-M. Kwon, M. J. Kim, S. Kang, and T. Kim, "Development of a high-storage-density hydrogen generator using solid-state NaBH4 as a hydrogen source for unmanned aerial vehicles," Applied Energy, vol. 251, p. 113331, 2019. 10/01/2019.
12. B. Lee, S. Kwon, P. Park, and K. Kim, "Active power management system for an unmanned aerial vehicle powered by solar cells, a fuel cell, and batteries," Aerospace & Electronic Systems IEEE Transactions on, vol. 50, no. 4, pp. 3167–3177, 2014.
13. H. Rezk, A. M. Nassef, M. A. Abdelkareem, A. H. Alami, and A. Fathy, "Comparison among various energy management strategies for reducing hydrogen consumption in a hybrid fuel cell/supercapacitor/battery system," International Journal of Hydrogen Energy, vol. 46, no. 8, pp. 6110–6126, 2021. 01/29/2021.
14. B. G. Gang and S. Kwon, "Design of an energy management technique for high endurance unmanned aerial vehicles powered by fuel and solar cell systems," International Journal of Hydrogen Energy, vol. 43, no. 20, pp. 9787–9796, 2018. 05/17/2018.
15. T. Chu et al., "Performance degradation and process engineering of the 10 kW proton exchange membrane fuel cell stack," Energy, vol. 219, p. 119623, 2021. 03/15/2021.
16. R. O. Stroman, M. W. Schuette, K. Swider-Lyons, J. A. Rodgers, and D. J. Edwards, "Liquid hydrogen fuel system design and demonstration in a small long endurance air vehicle," International Journal of Hydrogen Energy, vol. 39, no. 21, pp. 11279–11290, 2014. 07/15/2014.
17. W. Sun et al., "Improving cell performance and alleviating performance degradation by constructing a novel structure of membrane electrode assembly (MEA) of DMFCs," International Journal of Hydrogen Energy, vol. 44, no. 60, pp. 32231–32239, 2019. 12/06/2019.
18. J. Lee, S. Lee, D. Han, G. Gwak, and H. Ju, "Numerical modeling and simulations of active direct methanol fuel cell (DMFC) systems under various ambient temperatures and operating conditions," International Journal of Hydrogen Energy, vol. 42, no. 3, pp. 1736–1750, 2017. 01/19/2017.
19. Y. Zhao et al., "Recent progress on solid oxide fuel cell: Lowering temperature and utilizing non-hydrogen fuels," International Journal of Hydrogen Energy, vol. 38, no. 36, pp. 16498–16517, 2013. 12/13/2013.
20. A. Gong and D. Verstraete, "Fuel cell propulsion in small fixed-wing unmanned aerial vehicles: Current status and research needs," International Journal of Hydrogen Energy, vol. 42, no. 33, pp. 21311–21333, 2017. 08/17/2017.
21. Z. Ji, M. M. Rokni, J. Qin, S. Zhang, and P. Dong, "Performance and size optimization of the turbine-less engine integrated solid oxide fuel cells on unmanned aerial vehicles with long endurance," Applied Energy, vol. 299, p. 117301, 2021. 10/01/2021.
22. A. Baroutaji, T. Wilberforce, M. Ramadan, and A. G. Olabi, "Comprehensive investigation on hydrogen and fuel cell technology in the aviation and aerospace sectors," Renewable and Sustainable Energy Reviews, vol. 106, pp. 31–40, 2019. 05/01/2019.
23. H. Barthelemy, M. Weber, and F. Barbier, "Hydrogen storage: Recent improvements and industrial perspectives," International Journal of Hydrogen Energy, vol. 42, no. 11, pp. 7254–7262, 2017. 03/16/2017.

24. O. Z. Sharaf and M. F. Orhan, "An overview of fuel cell technology: Fundamentals and applications," Renewable and Sustainable Energy Reviews, vol. 32, pp. 810–853, 2014. 04/01/2014.
25. S. Wu et al., "A review of modified metal bipolar plates for proton exchange membrane fuel cells," International Journal of Hydrogen Energy, vol. 46, no. 12, pp. 8672–8701, 2021. 02/16/2021.
26. N. Li, X. Liu, B. Yu, L. Li, J. Xu, and Q. Tan, "Study on the environmental adaptability of lithium-ion battery powered UAV under extreme temperature conditions," Energy, vol. 219, p. 119481, 2021. 03/15/2021.
27. E. Özbek, G. Yalin, S. Ekici, and T. H. Karakoc, "Evaluation of design methodology, limitations, and iterations of a hydrogen fuelled hybrid fuel cell mini UAV," Energy, vol. 213, p. 118757, 2020. 12/15/2020.
28. A. Nishizawa, J. Kallo, O. Garrot, and J. Weiss-Ungethüm, "Fuel cell and Li-ion battery direct hybridization system for aircraft applications," Journal of Power Sources, vol. 222, pp. 294–300, 2013. 01/15/2013.
29. D. Verstraete, A. Gong, D. D. C. Lu, and J. L. Palmer, "Experimental investigation of the role of the battery in the AeroStack hybrid, fuel-cell-based propulsion system for small unmanned aircraft systems," International Journal of Hydrogen Energy, vol. 40, no. 3, pp. 1598–1606, 2015. 01/21/2015.
30. M. Dudek, P. Tomczyk, P. Wygonik, M. Korkosz, and B. Lis, "Hybrid Fuel Cell—Battery System as a Main Power Unit for Small Unmanned Aerial Vehicles (UAV)," International journal of electrochemical science, vol. 8, no. 6, pp. 8442–8463, 2013.
31. D. Verstraete, K. Lehmkuehler, A. Gong, J. R. Harvey, G. Brian, and J. L. Palmer, "Characterisation of a hybrid, fuel-cell-based propulsion system for small unmanned aircraft," Journal of Power Sources, vol. 250, pp. 204–211, 2014. 03/15/2014.
32. B. Wang, C. Wang, Q. Hu, L. Zhang, and Z. Wang, "Modeling the dynamic self-discharge effects of supercapacitors using a controlled current source based ladder equivalent circuit," Journal of Energy Storage, vol. 30, p. 101473, 2020. 08/01/2020.
33. S. N. Motapon, L. A. Dessaint, and K. Al-Haddad, "A Comparative Study of Energy Management Schemes for a Fuel-Cell Hybrid Emergency Power System of More-Electric Aircraft," *IEEE Transactions on Industrial Electronics,* vol. 61, no. 3, pp. 1320–1334, 2013.
34. A. Gong, R. Macneill, D. Verstraete, and J. L. Palmer, "Analysis of a Fuel-Cell/Battery/Supercapacitor Hybrid Propulsion System for a UAV using a Hardware-in-the-Loop Flight Simulator," in 2018 AIAA/IEEE Electric Aircraft Technologies Symposium, 2018.
35. S. Wang, D. Ma, M. Yang, L. Zhang, and G. Li, "Flight strategy optimization for high-altitude long-endurance solar-powered aircraft based on Gauss pseudo-spectral method," Chinese Journal of Aeronautics, vol. 32, no. 10, pp. 2286–2298, 2019. 10/01/2019.
36. B. Lee, P. Park, C. Kim, S. Yang, and S. Ahn, "Power managements of a hybrid electric propulsion system for UAVs," Journal of Mechanical Science & Technology, vol. 26, no. 8, pp. 2291–2299, 2012.

# Fuel Cell–Powered Passenger Aircrafts

**Tine Tomažič**

## 1 Introduction

The intent of the chapter is to inform the reader about the historic implementation of hydrogen as the fuel on board aircraft, by presenting several key evolutionary findings pertaining to the aircraft design from the late 1800s onwards. Efforts related to hydrogen of the twentieth century as performed by European, Russian and USA focal research are discussed, and future pathways related to technological and certification challenges are presented. Alongside high-level challenges, a few candidate fuel cell powertrain architectures for passenger aircraft will be presented and elaborated, followed by a commentary pertaining to water vapour atmospheric deposition emissions.

## 2 Historic Overview

### *2.1 First Era: Lighter Than Air—Hydrogen as Buoyancy Medium*

Flying with hydrogen present on board an aircraft dated back to 1852 when Henry Giffard first flew a hydrogen-filled airship in Paris. In this first application, hydrogen was not used as fuel, but rather only as a buoyancy medium. By 1872, Paul Haenlein designed a hydrogen-filled airship, which used the same hydrogen used for buoyancy also to power an internal combustion engine. Still, these airships were

T. Tomažič (✉)
Pipistrel Vertical Solutions, Ajdovščina, Slovenia
e-mail: tine.tomazic@pipistrel-aircraft.com



C. O. Colpan, A. Kovač (eds.), *Fuel Cell and Hydrogen Technologies in Aviation*, Sustainable Aviation, https://doi.org/10.1007/978-3-030-99018-3_4

not capable of round-trip flights, let alone as passenger transportation means. This changed in 1884, courtesy of Charles Renard and A.C. Krebs, showcasing the viability of the hydrogen-filled airship for commercial air transport, paving the way to the first commercial airship in 1900. The production of the rigid-construction LZ-1, built by Count Ferdinand von Zeppelin, began, and by 1916 five airships were operating commercially within Germany, completing 1600 flights and having carried 37,250 people without incident.

The next major development in airships occurred in 1928 with the LZ-127. This one airship alone flew over one million miles carrying more than 13,000 passengers and 118 tons of cargo over 590 flights in 9 years. Its derivative, the L-129, better known as the Hindenburg, was the second to last in the LZ-line. It made history when it caught on fire and crashed in Lakehurst, NJ in May 1937. Thirty-five of the 97 people on board lost their lives, with most of the fatalities caused after passengers jumped from the airship in distress. Contrary to popular belief, the intense fire seen on Hindenburg accident photographs and videos, which still to date create ripples in public perception of hydrogen-based flight, is not hydrogen itself seen burning, but rather the fabric and resin that was forming the airship's skin. Hydrogen burns with an invisible flame. Even with the inclusion of this one tragedy, commercial hydrogen airships had an impressive safety record, despite their flimsy construction.

## 2.2 *Second Era: Hydrogen as a Fuel for Burning*

With no apparent progress using hydrogen for airborne applications during the WW2 period, efforts restarted in 1954 with Lockheed proposing a design for an aircraft capable of cruising at 2.5 Mach at 100,000 ft., powered by hydrogen fuel burned inside jet turbines. The CL-400 was never built, and its original purpose was filled by the A-12—the lack of hydrogen infrastructure again proves to be a recurring theme and reason for the termination of hydrogen aircraft programs. It is little known that despite CL-400 program cancellation, the technology necessary to prove the feasibility of a plane burning hydrogen as fuel was indeed developed.

The testing performed with a modified B-57B airplane, courtesy of NACA Lewis facility also showed that liquid hydrogen could be handled as easily and safely as hydrocarbon fuel [1]. One of the two engines of this modified airplane was run entirely on hydrogen, with the other remaining fuelled with hydrocarbon. The programme reported smooth and reliable operation of both engines with the aircraft operating at up to 50,000 ft. and 0.75 Mach. The 1950s were an important period for validating the feasibility of hydrogen aircraft technology and made the United States a world leader in this area of research.

In the heat of the space race, hydrogen in combination with liquid oxygen was first used in a rocket engine in 1963. There, hydrogen also replaced the lubricants and acted as the heat-sink for the rocket. The Saturn V later used to launch the United States Apollo modules used over 95,000 kg of liquid hydrogen. There was never a failure of the hydrogen-fuelled rocket engines, and based on this success in 1973, NASA began to study hydrogen-powered aircraft in great detail.

The studies were extremely detailed and covered every component part, this time also including the airport infrastructure necessary to fuel the aircraft. The results of these studies suggested that liquid hydrogen was not only a feasible choice for air mobility but that it was the preferred alternative to hydrocarbon fuel, largely because of the increased energy density of the fuel itself, and its abundant availability. Brewer, one of the authors of the NASA studies comparing synthetic jet fuel, liquid methane and liquid hydrogen aircraft states, 'The LH2 design is superior in nearly every basis of comparison. Its gross weight, fuel weight, and operating empty weight are all significantly less' [1]. At that time, coinciding with the conclusion of the study, the price of oil was so low that a managemental decision was made to continue researching applications with hydrocarbon fuels in favour of hydrogen application, despite its performance advantages.

NASA studies showcased characteristic design proposals for passenger carrying hydrogen-powered aircraft, clearly depicting the need for a slightly larger and longer fuselage for hydrogen aircraft. The location of the hydrogen tanks was always inside the fuselage of the aircraft, opposing the typical design choice of hydrocarbon-powered aircraft that carry fuel in their wings. The main rationale for tank placement into the fuselage is because liquid hydrogen needs to be maintained at cryogenic temperatures, where to minimize heat transfer, the tanks benefit from having a minimal surface area to volume ratio. To achieve this, the best geometric shape is a sphere, with the cylinder being the next best option. Concurrently, these shapes also cater best gravimetric efficiency of the tank as the pressure vessel. Therefore, hydrogen aircraft tends to have large cylindrical tanks placed in the fuselage, reducing deck space typically used to fit the passenger cabin and/or cargo.

One of the immediate downsides of integrating hydrogen tanks into airframes is associated with its significantly lower volumetric energy density—hydrogen hence requires larger integration volumes than liquid carbohydrate fuels. Also, when liquid hydrogen is used as fuel, it needs to be laboriously insulated and kept cool in order to stay in liquid form—saving space on board the airplane. TU-155 engineers addressed both these issues already from the beginning of the developmental programme. A TU-154, registered as CCCP-85035, was modified by fitting a hydrogen tank in the rear of the passenger cabin—taking approximately one-third of the space there. More than 30 additional systems were integrated into the aircraft so that it could be propelled with hydrogen, forming the TU-155. On 15 April 1988, it became the world's first hydrogen-powered airliner to take flight.

This achievement almost coincided with the beginning of the Cryoplane project. The Cryoplane project was built on top of the work NASA had done in the 1970s in order to validate technological capabilities of integrating and operating aircraft to use liquid hydrogen as fuel. Contrary to TU-155 tank installation, which was inside the fuselage, NASA concluded that the tank should not be internal and suggested it be designed as part of the aircraft's structure to help reduce overall aircraft weight.

The Cryoplane project, which began in 1990, was a joint project between Daimler-Benz Aerospace Airbus and Tupolev with the aim of developing a passenger aircraft to be propelled by liquid hydrogen instead of jet fuel. The study was initiated due to the economics of rising fuel prices and jet fuel's impact on air pollution, greenhouse gas emissions and ozone depletion [2]. The first phase of the project, from 1990 to 1993, focused on a feasibility study of a modified A310 converted to run on liquid hydrogen. The project considered a modification of the A310 that differed from the 1970s NASA conceptual designs in several important aspects. First, the tanks were not located fore and aft of the cabin, but were instead located above the passengers. This design is proved to be suboptimal for multiple reasons. Most notably, such design adds drag due to a larger frontal cross-section. In addition to adding drag, the design also contributes a substantial weight over the 1970s NASA designs as reinforcements are needed above the passenger cabin in order to carry the weight of the installed tank. The conclusion was quite clear: modifying an aircraft designed for jet fuel to run on liquid hydrogen fails to take advantage of liquid hydrogen's unique properties, a clean-sheet design of the aircraft is necessary.

Following this theoretical phase work inside project cyroplane also included tests which were performed on various technology building blocks—from 1992 to 1996, the Euro-Quebec Hydro-Hydrogen Pilot Project combustion chamber tests took place—from 1994 to 1999, Tupolev, Airbus and Air Liquide collaborated on liquid hydrogen tank tests—from 1995 to 1998, Germany and Russia collaborated on a demonstrator aircraft based on a Dornier 328, and from 2000 to 2002, a systems analysis of liquid hydrogen aircraft implementation was conducted. The results of the systems analysis illustrated the feasibility of hydrogen aircraft technology.

During the course of Cryoplane project, several advances also took place in the United States—some with importance still today. In 1988, an aviation enthusiast flew the first powered flight in a small single engine using only hydrogen as fuel, and in 1996, NASA selected Lockheed Martin's Skunk Works to design a Single-Stage-to-Orbit vehicle. The X-33 was to be the prototype for this concept—a rocket-based vehicle, however featuring cryogenic hydrogen storage and full vehicle reusability. This makes the nature of X-33 technologies similar to that needed and applicable to hydrogen-powered passenger aircraft. Unfortunately, the programme was cancelled in 2001 when the aircraft was already 85% complete. The primary reason for the cancellation was the weight of the composite liquid hydrogen tank exceeding requirements. The engineers had originally suggested an aluminium-lithium alloy liquid hydrogen tank similar to the space shuttle since the technology for a composite tank was not mature. An important take-away was that despite the composite tank being lighter in the skins, it was heavier in the joints than the aluminium tank.

## 2.3 *Third Era: Hydrogen as a Fuel for Fuel Cell–Based Propulsion*

In 2007, the Georgia Institute of Technology built a demonstrator aircraft to test the performance of a hydrogen fuel cell aircraft. The tests provided data for the comparison of fuel cell aircraft performance to that of conventional aircraft (Bradley, 2007).

In 2008, Boeing constructed the first hydrogen manned aircraft to use a fuel cell. The plane was capable of flying for 45 min, but the tests were for only half that amount of time. This manned aircraft can be seen below.

In 2009, the ENFICA-FC consortium led by University of Turin flew the Rapid 200-FC, a PEM fuel cell–powered manned airplane, using a 20 kW fuel cell and compressed, gaseous, hydrogen at 350 bar as fuel.

In the same year, the German Aerospace Center's Antares DLR-H2 became the first manned research airplane with the capability to take off solely using fuel cell power. Based on the power glider Antares 20E, it has been developed to flexibly and cost-efficiently test airborne fuel cell systems. The airplane was able to cover distances in excess of 700 km.

In 2016, the DLR-led partnership flew Hy4, the world's first four-seat hydrogen fuel cell–powered airplane, which was further improved in 2020 under EU project MAHEPA, achieving redundancy of powertrain and fuel system, as well as multi-hour endurance.

In 2020, Zeroavia flew their fuel cell–powered demonstrator. Installation and test of the ZA250 hydrogen–electric powertrain in a six-seat Piper Malibu culminated in an 8-min first flight of the hydrogen-electric Malibu in September that year. Later details released describe the powertrain as being 250 kw battery-hydrogen hybrid with 100 kw from a PowerCell MS-100 hydrogen fuel cell. A larger 19-seat size demonstrator to facilitate a 600 kW FC powertrain was announced by the company in 2021.

2021 brings announcements by Universal Hydrogen and Deutsche Aircraft and H2Fly to develop megawatt-class fuel cell powertrains to be fitted to Bombardier's DHC-8/400 and Dornier 328 respectfully by mid-decade.

## 3 Fuel Cell Powertrain Architectures

For the purpose of this chapter, only all-electric (pure fuel-cell or battery/fuel-cell) hybrids are discussed.

Many possible architectures can be facilitated in terms of power delivery, which would typically be linked to the aircraft concept where such powertrains would be installed. Most variability in these application-specific architectures would be observed in:

- Modularity/redundancy of the fuel cell system (number of fuel cells and how they interconnect)
- Set-up of hydrogen storage and delivery systems (number of fuel tanks and execution of fuel supply)
- Interaction with batteries and voltage stabilization/control

While initially single-string architecture with passive voltage stabilization when hybridized with batteries were employed, modern architecture takes advantage of state-of-the-art power densities of DC/DC power electronics solutions to achieve advanced fuel cell system functionalities, including in-flight battery recharge and improved energy density, thermodynamic efficiency and life cycle durability.

### 3.1 Contrast to $H_2$ Burn

Most historic endeavours related to hydrogen-powered flight resorted to hydrogen burning in a turbine (or rocket engine), not because of better efficiencies, but because of much higher power densities of combustion engines relative to fuel cells and much easier thermodynamic management of the complete powertrain. With the power density of fuel cell stacks surpassing 4500 W/kg (2021) and virtually double thermodynamic efficiency relative to medium size (<25 MW) turbine engines, low/medium temperature PEM fuel cells are becoming viable technology for powering regional (~50 seats) airplanes. One of the greatest challenges however remains the heat management (cooling) of the fuel cells, which exhaust approximately an equivalent amount of their electrical power output in low-quality heat.

### 3.2 Fuel Systems

Fuel systems have the function of delivering hydrogen from their storage vessel to the fuel cell. In principle, hydrogen distribution systems will be divided by pressure (low pressure vs. high pressure fuel systems) and state of hydrogen being distributed (liquid vs. gaseous). The latter is most often associated with operational temperatures, although there may be variants of fuel systems that are delivering cryogenic, but still gaseous hydrogen to the fuel cell.

The fuel cell itself requires gaseous hydrogen to function; therefore, designers of fuel systems must decide where the state transition between liquid and gaseous form hydrogen will occur, in case liquid hydrogen is used as fuel/energy storage.

There are three intrinsic possibilities:

(a) Hydrogen transitions from liquid to gaseous form at the hydrogen tank with or without active boil-off management.
(b) Hydrogen transitions from liquid to gaseous form just before entering the fuel cell.
(c) Hydrogen transitions from liquid to gaseous form in a dedicated element of the fuel system, which would be dedicated to (at least partially) managing boil off.

Given that all gaseous hydrogen storage and fuel delivery systems, at the current state of the art, are not viable for scale-up to airliner size applications, the inclusion of liquid hydrogen as fuel also begins to present opportunities for the problematic heat management of the fuel cell stacks.

## 4 Heat Management

A fuel cell system typically has three heat output mechanisms:

– Hot cooling water (90% of total)

– Cathode exhaust (~9% of total)
– Un-reacted hydrogen (<~1% of total)

Aircraft designers can consider potential utilization of generated heat instead of simply looking for ways to reject it into the atmosphere. Candidate on board heat uses in a fuel cell–powered passenger aircraft may include:

- Liquid hydrogen preheating
- Absorption by airplane's cooling system
  - Anti/de-icing
  - Hot water
  - Food preparation

Analogously, the architectures of heat management will influence the overall fuel cell powertrain implementation. Three primary differences in heat management system configuration and methods are presented.

## 4.1 *Air-Cooled (Complete Rejection of Heat into the Atmosphere)*

This execution is architecturally the simplest (Fig. 1), however also requires the most consideration of cooling-drag and size of heat-exchange elements of the fuel cell system. Evidently, the re-use of heat output of the fuel cell is none, yielding a thermal reuse factor Tea = 0.

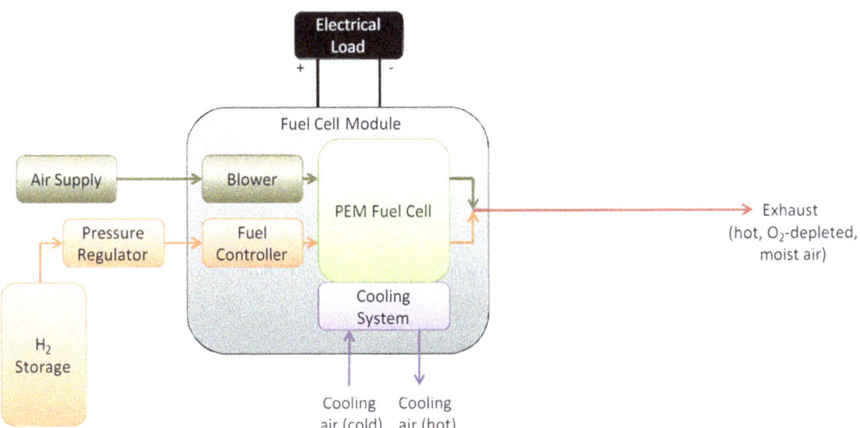


**Fig. 1** Notional air-cooled fuel cell power module architecture

## 4.2 *Liquid-Cooled, with Maximum Water Recovery with Hot Water Availability*

This variant rejects fuel cell heat into water (liquid) which is thereafter used to provide different utility functions (Fig. 2). A large portion of the waste heat is still emitted into the atmosphere. Aircraft designers should consider trade-offs of such heat management architectures.

## 4.3 *Cryogenic Fuel Cooled*

This heat management architecture is directly opposing to the all air-cooled variant, as all of the heat the PEM fuel cell is generated is absorbed into fuel (Fig. 3). This utilization may not always be possible, or practical; however, it represents the higher performance options as it eliminates the notion of cooling drag.

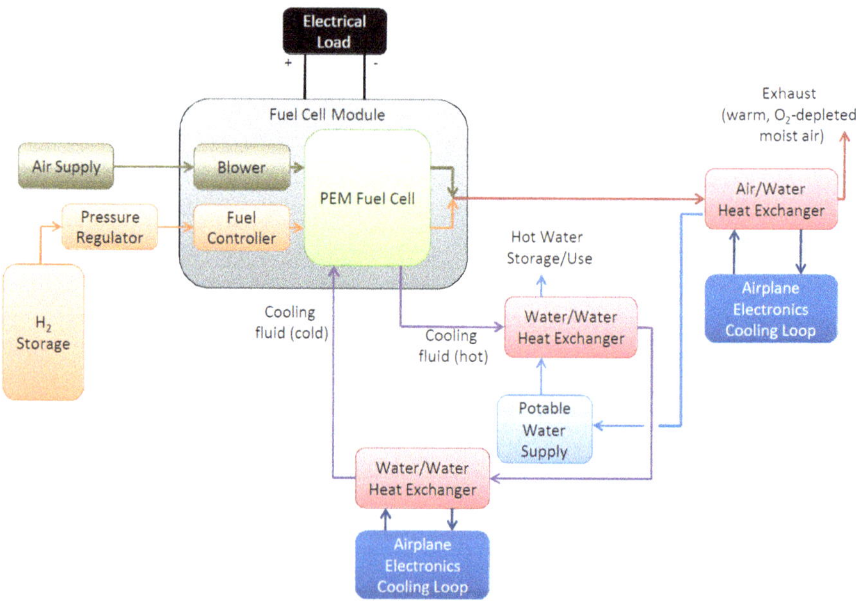


**Fig. 2** Notional liquid-cooled fuel cell power module architecture, with thermal recovery

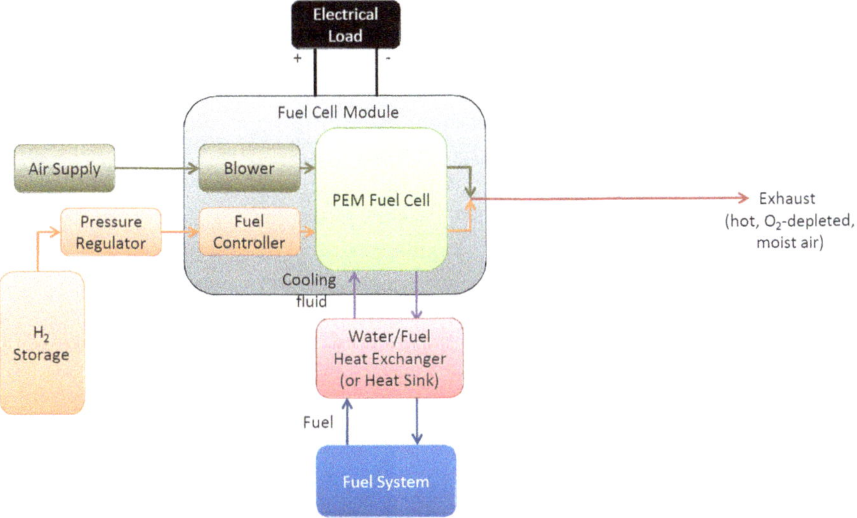


**Fig. 3** Notional cryo-cooled fuel cell power module architecture

## 5 The Balance-of-Plant

Balance-of-plant is a term generally used in the context of fuel cell systems to refer to all the supporting components and auxiliary systems surrounding the fuel cell stack. In essence, the balance-of-plant provides the functions of

- Fuel processing
- Air intake–air exhaust
- Fuel cell monitoring
- Low-level fuel cell stack control
- Electrical power conversion / output management

Components such as heat exchangers, humidifiers, de-ionization filters, cell monitoring modules, DC/DC voltage stabilization or conversion circuitry, etc. are also regarded as part of the balance-of-plant, which mass and volume often surpass that of the fuel cell stack alone (Fig. 4).

Depending on the fuel cell powertrain architecture, one set of balance-of-plant components may handle one or more fuel cell stacks. By volume and weight, balance-of-plant may be greater than the "naked" fuel cell stack itself; therefore, careful packaging and placement considerations need to be made at the aircraft design stage already. It is the balance of plant that differentiates how the FC-powered aircraft is designed and configured, rather than the fuel cell stack itself.

**Fig. 4** Balance-of-plant of Hy4 fuel cell–powered airplane nested around fuel cell stacks

## 6 Airframe Integration

Besides Cryoplane and Airbus ZEROe concepts (2020), which mostly focus on regional and single-aisle missions, the German design house Bauhaus Luftfahrt has conducted preliminary design studies on concept Hyliner. The operational airspeed was set to Mach 0.7, allowing for significant energy savings. Concurrently, the increase of capacity to 400 passengers permitted to keep productivity at the same level.

For the resulting aircraft concept Hyliner (2.0), liquid hydrogen (LH2) is chosen as the energy source, which is burned in gas turbines but requires significantly larger volumes on board for storage (up to four times) compared to using compressed hydrogen. Since conventional wing tanks neither have the required storage volume nor can meet the extended requirements for thermal insulation and compressive strength, fuselage tanks are used. Such an LH2 aircraft can, due to its reduced weight despite the higher volumes, be designed energy equivalent to a conventional aircraft of the same technology level. As part of the project, the fuselage size is enlarged by the extra space and the service options of a third, passenger-friendly deck, leading to an increase in energy consumption of 9%, concluding with a powerful statement which outlines the design space of future hydrogen-powered aviation, including aircraft powered by fuel cells: 'While the energy consumption of the Hyliner (2.0) is 9% higher compared to the conventional reference aircraft with the same technologies integrated, the combustion of hydrogen rather than kerosene offers a possible reduction of the climate impact on long range operations'.

# 7 Safety and Certification Challenges

As of 2021, there are no regulations or standards in place providing a clear path to certification for hydrogen-powered heavier-than-air aircraft. That said, two working groups are making efforts to provide standards and means-of-compliance materials.

## *7.1 EUROCAE/SAE*

The purpose of the working group WG-80, operating jointly with the SAE AE-7A committee and having 80+ members, is the development of minimum standards to support qualification and certification of hydrogen fuel cell systems in various intended applications for civil large (Part 25) aircraft. Objectives and Scope of WG-80, which was founded in December 2008 and operates as a joint EUROCAE/SAE working group with SAE AE-7A are to develop guidelines to support the use of hydrogen (beginning with gaseous one) and oxygen supplied PEM fuel cell systems for civil aircraft applications. The group is particularly focused on the safety and functionality of the systems, including balance of plant provision; however, it does not deal with performance requirements (such as power) and reliability—these are out of scope of this standardization activity. In 2013, the AIR6464/ED-219 'Aircraft Fuel Cell Safety' guidelines) was published as a result of the early. Further in 2019, the group has published AIR7765/ER-20, a comprehensive document for 'decision makers' on hydrogen, its application and benefits for aircraft, providing insight about why and how to use hydrogen and fuel cells in aviation.

The effort has begun by encompassing three target non-propulsive applications: a fuel cell–powered galley, a medical application and a fuel cell emergency power system to replace a ram air turbine. These are also representative examples of non-propulsive use of fuel cells on board passenger aircraft. AS6845/ED-245 'Installation of fuel cell systems on large civil aircraft' followed in 2017, detailing specification provided for the three mentioned PEM systems.

As of 2016, the focus has shifted to PEM fuel cell propulsion in combination with fuel systems and considering LH2 fuel storage, fuel distribution. Solid Oxide Fuel cells have not yet been discussed. AS6679, anticipated for publication in 2021, describes LH2 general properties and system definition, as well as specifies critical requirements for safe use of liquid hydrogen onboard aircraft and on ground in terms of: operation, storage and distribution; maintenance; safety; qualification; installation and certification.

### 7.2 ASTM

Less applicable to large passenger aircraft, the ASTM's working group F44.40, tasked with the development of industry consensus standards for General Aviation airplanes, has taken on board the responsibility of developing means-of-compliance performance based standards for hydrogen fuel systems and fuel-cell integration to Part-23 airplanes. One must not neglect the fact that Level-4 (commuter sized) Part-23 airplanes approach safety/reliability levels very close to those of Part-25, therefore some overlap with EUROCAE/SAE's work is to be expected and diligently considered by aircraft designers, especially for smaller passenger fuel cell–powered aircraft.

### 7.3 EASA

EASA's 2021 publication of Special Condition E-19 does not contain elements of hydrogen-based fuel cell propulsion.

### 7.4 FAA Publications

The FAA has invited EUROCAE/SAE members to participate in the Energy Supply Device Aviation Rulemaking Committee (2015), with main focus on fuel cell systems. The report was finished and published by the FAA in April 2019: it includes recommendations for airworthiness and operational rules, standards, as well as identifies the need for change of existing requirements and need for guidance publications.

## 8 Non-$CO_2$ Emissions and Their Impact

While hydrogen-fuel cell combustion practically eliminates $CO_2$ and $NO_x$ emissions, given that no combustion occurs, the emission scenarios related to water vapour / contrails may not be disregarded. Contrail-cirrus and ozone formation have recently been assessed by Volger Grewe et al. [3]:

> The formation of persistent contrails-cirrus depends on aircraft and fuel parameters as well as atmospheric conditions, as the propensity of contrail formation is higher in the cold and saturated atmosphere. Contrail-cirrus influence the incoming solar radiation and the outgoing infrared radiation emitted by the Earth and its atmosphere. The net change, the radiative forcing (RF), is on average positive and hence contrail-cirrus act to warm the climate.

Due to improved modelling of contrail microphysics, the authors were able to conclude that:

> 'the increase in transport volume leads to an increase in the overall climate impact from aviation, which also increases the relative importance of $CO_2$ (25% in 2005 Base to 39% in 2100 CurTec), even if aviation net $CO_2$ emissions are regulated and capped to 2020 values. The increase in fuel efficiency of aviation technologies at a current rate decreases the overall climate impact, especially for $CO_2$ and $NO_x$. By this, it mainly reduces the relative contribution of $NO_x$ (41–16%). The introduction of the CORSIA scheme further reduces the climate impact of $CO_2$ emissions and that increases the relative importance of contrail-cirrus and $NO_x$. The technological measures from FP2050 have a similar reduction efficiency for $CO_2$ as CORSIA, however, the strong measures for $NO_x$ largely reduce the overall climate impact so that the remaining climate impact from aviation is due to $CO_2$ (50 60%) and contrail cirrus (around 30%)'. Which points to favourable use of hydrogen fuel cell propulsion for passenger aircraft—given that $CO_2$ and $NO_x$ are not even emitted in the first place.

## 9 Concluding Remarks

Perhaps the most telling publication linked to the strategic development of fuel cell propelled passenger airplanes is the Clean Aviation Strategic Research Innovation Agenda [4], not only setting clear goals for fuel cell powertrains intended for passenger aircraft to be developed and matured by the end of the decade, but it also outlines clearly the pathway towards this goal. Adjacent industries to aviation have matured fuel cell stacks to performance levels beyond 4000 Watts per kg (plate-to-plate, before BOP) in 2021, which unlocks sizable aircraft, including ones beyond 50 seats in size. An appropriate balance of plant implementation and aircraft specific integration remain a challenge requiring transversal knowledge of thermo- and aero-dynamics, cryogenics and systems safety engineering. Associated ground infrastructure readiness will need to be facilitated concurrently with the development of airborne equipment. The current realistic outlook expects fuel-cell powertrains for aviation of adequate, megawatt, power level to arrive at TRL 6 maturity by 2026–2028 with an entry into service post 2030.

## References


1. Brewer (1991): Brewer, G.D.: "Hydrogen Aircraft Technology", CRC Press, 1991
2. Schmidtchen, 1998: U. Schmidtchen et al: Hydrogen aircraft and airport safety, Renewable and Sustainable Energy Reviews, Volume 1, Issue 4, December 1997, Pages 239–269
3. Evaluating the climate impact of aviation emission scenarios towards the Paris agreement including COVID-19 effect, Volger Grewe et al, Nature Communications (2021)12-3841
4. Clean Aviation SRIA https://www.clean-aviation.eu/files/Clean_Aviation_SRIA_R1_for_public__consultation.pdf, May 2020

# Energy Management Strategies in a Fuel Cell–Powered Aircraft

**Pedro Muñoz, Enrico Cestino, and Gabriel Correa**

## 1 Introduction

With growing concerns about aviation-related greenhouse gas emissions, noise pollution near airports, and fossil fuel scarcity, the global aviation industry is undergoing a major paradigm shift [1]. Aviation is one of the fastest-growing sources of greenhouse gas emissions, accounting for 3% of global $CO_2$ emissions [2] and 3.6% of emissions in Europe [3].

However, the aviation sector is undoubtedly one of the most challenging sectors to decarbonize. More importantly, when compared to the ground mobility sectors, the mass/weight restrictions of the aircraft exclude many potential low-emission technologies due to the need for very high energy mass density.

As in the automotive industry, electrification of aviation is also an inevitable trend, which has seen many emerging electric aircraft technologies [4]. However, current battery technology is an example where, despite major advances in the increase of power density, the state of the art is still several factors below of what would be needed for major adoption in large commercial aircraft.

Rapidly emerging hydrogen-based technologies can be the base to the shift toward high-reliability and low-maintenance electric propulsion systems for general aviation and small commuter aircraft, as well as electrical system replacement (emergency power, cabin power, auxiliary power unit, anti-icing system, landing

P. Muñoz · G. Correa (✉)
CREAS, Facultad de Ciencias Exactas y Naturales, Universidad Nacional de Catamarca, CONICET, San Fernando del Valle de Catamarca, Argentina
e-mail: correa.gabriel@conicet.gov.ar

E. Cestino
Department of Mechanical and Aerospace Engineering, Politecnico di Torino, Torino, Italy
e-mail: enrico.cestino@polito.it

 
C. O. Colpan, A. Kovač (eds.), *Fuel Cell and Hydrogen Technologies in Aviation*, Sustainable Aviation, https://doi.org/10.1007/978-3-030-99018-3_5

gear retraction, etc.) for larger transport aircraft. Such power sources have several potential advantages, ranging from environmental aspects (emissions and noise) to improved performance and operability.

Replacing internal combustion engines with electric engines can drastically reduce the emissions ($CO_2$, water and smaller amounts of $SO_2$, CO, $NO_x$, and unused hydrocarbons) that are responsible, directly or indirectly, for the greenhouse effect and the air pollution.

Nevertheless, an aircraft powered only by a fuel cell system (FCS) would have serious limitations, as its slow dynamic response may not be suited for situations where the power demand changes drastically, and the peak power outputs take a toll on the fuel cell durability. These situations make necessary hybridization with other power sources [5, 6].

The age of zero-emission aviation is limited by the available battery performances in terms of energy efficiency and specific power. Romeo et al. [7] show that to fly a plane for 1000 km with a typical battery ratio of 0.4–0.5 and aerodynamic efficiency from 10 to 15, an energy density of about 500 Wh/kg is required but the actual Li-ion battery energy density is limited to about 200 Wh/kg. Hydrogen has the necessary energy density, but the main barrier is the specific power limited to about 1000 W/kg with an aviation requirement set to about 1500 W/kg. A hybrid battery plus hydrogen fuel cell system could represent a short-term solution especially for electric vertical take-off and landing (eVTOL) aircraft where high specific power is required during take-off and landing.

Another critical factor is the charging time or how quickly the battery can be returned to its fully charged state. Tesla has already shown the efficiency of rapid chargers by achieving an 80% battery charge within only 30 min. Nevertheless, rapid recharging can significantly reduce the battery lifespan, which consequently means a higher risk solution when we talk about air mobility.

Sufficient charging speed is essential in an eVTOL application as a city air taxi vehicle for which low operating times between flights are required. Furthermore, the high vehicle utilization rate poses a critical challenge to battery cycle life. For these reasons, a power management optimization system, capable of guaranteeing a sufficient residual charge at the time of landing, can be of considerable advantage for these innovative mobility systems based on battery and fuel cell hybridization.

Given that the hybridization of power sources for the fuel cell–powered aircraft is regarded as a good design that allows to reduce the FCS weight by decreasing the fuel cell (FC) maximum power, three questions arise: how to size each power source, when to use each power source, and in which proportion if they are to be used simultaneously. To answer these questions, it is necessary to first layout the objectives, i.e., minimize the hydrogen consumption, minimize the equivalent energy consumption (hydrogen and battery), extend the life of the fuel cell and the battery, maximize the traveled distance, reduce operation costs, etc. To meet these objectives, there is a myriad of minimization (or maximization) methods available in the literature for different types of problems, under different constraints [8], and the algorithms are widely available for different programming languages [9, 10]. The description of these methods is beyond the scope of this chapter, and it is a book on its own.

Energy management strategies (EMSs) can also be obtained by rule-based methods. These methods are more straightforward and easy to understand because they come from engineering intuition. However, they often lack optimality or cycle-beating. On the other hand, optimization-based EMSs require the cycle to be known a priori. This represents a major problem when the power requirements are unpredictable, as in the case of a car. In aircrafts, this obstacle can be surpassed as the velocity and height profiles can be known a priori from the flight plan.

The literature on EMSs is abundant for the fuel cell hybrid cars or buses [11–16] but is scarce in the electric aircraft's field. Doff-Sotta et al. [17] use model predictive control on a mathematical model of a BAe 146 aircraft powered by a gas turbine and a battery to minimize the fuel consumption under battery restrictions, showing up to 3.8% reduction in fuel consumption using optimized strategies. Zhao et al. [18] propose to manage the energy from two power sources from an emergency power unit with two strategies, one that maximizes the supercapacitors energy output and one that minimizes the equivalent fuel consumption of the fuel cell and the batteries, and compare nine meta-heuristic optimization techniques, measuring the performance of the obtained EMS in seven dimensions: battery state of charge difference, hydrogen consumption, overall efficiency, fuel cell stress, battery stress, supercapacitor stress, and execution time. They conclude that the moth swarm algorithm gives the best results for a weighted average of all the measured dimensions while the electromagnetic field optimization method yields the most significant hydrogen consumption reduction. Li et al. [19] do a multi-objective parameter optimization to define the optimal sizing of the energy storage systems and the hyper-parameters in the power controller while minimizing the hydrogen cost of the voyage and the aging cost of the cell management minimizing the hydrogen cost of the voyage and the aging cost of the cell. In that work, the authors use the mathematical model of X-57 Maxwell with hybrid fuel cell and battery propulsion, and the optimization yields a 16.7% reduction in hydrogen consumption with a 66.4% fuel cell life loss mitigation.

The scarcity of scientific literature available in this area combined with the favorable results from the previous works indicates that this is a promising research topic where there is much to be done. In this chapter, the state of the art of energy management strategies will be described and an example of optimization of power management to attain the maximum flight time will be given for two different aircraft powertrains.

## 2 Optimized Management Strategies of Two Very Light Aircraft Configurations

To demonstrate the benefits of defining the energy management strategies and sizing via optimization, two different types of reference aircraft were considered: (1) a general aviation (GA) aircraft, propeller-driven and powered by hydrogen fuel cells and batteries and (2) an eVTOL configuration for urban mobility based on a distributed propulsion consisting of 36 ducted fans and powered by both battery and hydrogen fuel cell system. Both the aircrafts belong to the very light aircraft category but differ in the power requirement especially in the take-off and landing phases.

## 2.1 *Aircraft Model*

The aircraft powertrain consist of a fuel cell system as its main power source, and a lithium-ion battery that is used mainly as an auxiliary power source. The FCS and the battery are connected in parallel, using DC/DC converters, to an electric bus that feeds the electric motors. The powertrain components layout can be seen in Fig. 1.

### Fuel Cell Model

Due to the high computation time needed to run a complete FCS dynamic model, the optimization process of each aircraft mission is carried out using a simplified model. It is necessary to point out that the optimization algorithm takes approximately 100 iterations to achieve convergence, and each iteration requires N evaluations of the objective function, where N is at least twice the number of elements in the optimization vector and the evaluation of the objective function requires running the FCS model at least one time. Lastly, the dynamic model can take up to 1200 s to be solved using Simulink on a desktop computer. For the optimization process, a map of the FCS hydrogen consumption was used due to the speed and ease of calculation. The consumption map was made using a validated proton-exchange membrane fuel cell stack and balance of plant dynamic model [20, 21]. The latter model was developed in Simulink and predicts the power output, temperature and hydrogen consumption taking into account the FCS main electrochemical, fluid-dynamic, and thermal processes. The balance of plant includes the cooling devices, the compressor, and the water management system. The model was validated (see Ref. [22]) with the experimental data of a demonstration session carried out within the ENFICA-FC project [7].

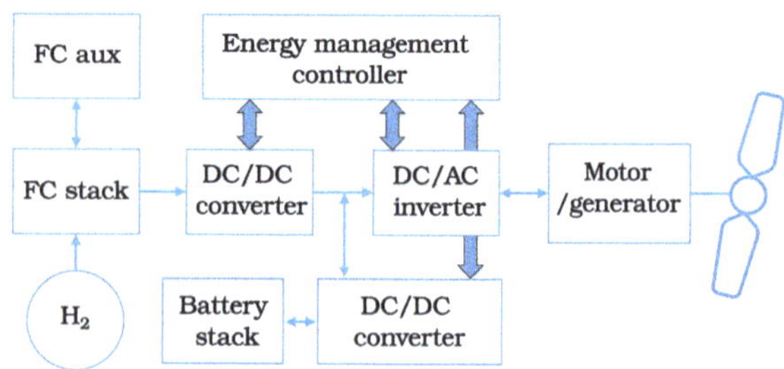


**Fig. 1** Schematic of the aircraft powertrain components

## Battery Model

The Li-ion battery model used in this work was developed by the authors in Ref. [23]. The battery is modeled from the charge and discharge curves of the cell voltage vs. state of charge (SOC) at different currents and different temperatures, obtaining a semi-empiric and quasi-static model that takes into account the voltage variation due to the battery temperature. Its inputs are the room temperature and the required power, and the initial values are the SOC and the battery temperature.

The charge and discharge curves of the battery form a surface that determines the voltage for every power requirement at a given SOC and temperature. Lastly, the current values are determined from the working voltage and the required power. Further details of the battery model can be found in Ref. [23]. The battery is operated in the range of 90% ≥ SOC ≥ 20% to avoid overcharges and overdischarges.

Since the power output from the FCS can be significantly different from that of the command signal due to its response time, the battery power is defined as the difference from the requested power and the power delivered by the FCS in order to fulfill the power requirements during the flight, as shown in Eq. (1).

$$P_{\mathrm{BAT}} = P_{req} - P_{FC\_\mathrm{out}} \tag{1}$$

## General Aviation Powertrain

The first reference configuration is a typical electric aircraft such as the Rapid 200FC [7, 21]. The typical aircraft mission consists of a taxi, a ground roll phase, a transition to climb and a climb to cruise altitude followed by the cruise condition. At the end, a descent and a landing phase completes the mission. For the purpose of this article, a Taxi phase with a duration of 2 min is considered at a speed of about 27 km/h with a constant power requirement of 11 kW.

After the Taxi phase, a take-off phase starts, including the ground roll, transition, and climb. During the ground roll phase, the traction forces on the aircraft are the thrust, aerodynamic drag, and friction on the wheels described by a friction coefficient $\mu$. The ground roll length can be calculated from:

$$S_{\mathrm{G}} = \frac{1}{2gK_{\mathrm{A}}} \ln\left(\frac{K_{\mathrm{T}} + K_{\mathrm{A}}V_{\mathrm{f}}^{2}}{K_{\mathrm{T}} + K_{\mathrm{A}}V_{\mathrm{i}}^{2}}\right) \tag{2}$$

where $V_{\mathrm{i}} = 0$ and $V_{\mathrm{f}} = 1.1V_{\mathrm{ST}}$ and:

$$K_{\mathrm{A}} = \frac{\rho}{2(W/S)}\left(\mu C_{\mathrm{L}} - C_{\mathrm{D0}} - KC_{\mathrm{L}}^{2}\right) \tag{3}$$

$$K_{\mathrm{T}} = \frac{T_{\mathrm{max}}}{W} - \mu \tag{4}$$

During the take-off, the aircraft aerodynamic is characterized by a 10° deflected flap configuration in which the quadratic aerodynamic drag polar approximation is estimated by:

$$C_{\mathrm{D}} = 0.05 + 0.039 \cdot C_{\mathrm{L}}^{2} \tag{5}$$

An $S_{\mathrm{G}} = 157$ m is computed considering a constant thrust of 1100 N, a friction coefficient $\mu = 0.03$, a stall speed of $V_{\mathrm{ST}} = 72$ km/h corresponding to a $C_{\mathrm{L_{max}}} = 1.84$. The ground roll time is about 14 s, and the power required is around 32 kW, assuming a total propulsive efficiency of $\eta_{\mathrm{P}} = 0.75$.

During the rotation and transition phases, the aircraft accelerates from take-off speed ($1.1V_{\mathrm{ST}}$) to climb speed ($1.2V_{\mathrm{ST}}$). The average vertical acceleration can be found to be around 1.2 m/s$^2$ as in Ref. [24]. The transition distance will be determined at the end, and the climb to the obstacle starts until the clearance is obtained at about 15 m. The Power required for this phase has been computed to be 38 kW. Finally, climb to cruise altitude starts, and power to climb is evaluated as:

$$P_{\mathrm{climb}} = P_{\mathrm{cruise}} + P_{\mathrm{vertical}} = \left(V_{\mathrm{c}} \cdot D + RC \cdot W\right)\frac{1}{\eta_{\mathrm{P}}} \tag{6}$$

During the climb, the pilot retracts the flaps, and the aircraft assumes the same aerodynamic configuration of the cruise where a new drag polar is activated:

$$C_{\mathrm{D}} = 0.015 + 0.065 \cdot C_{\mathrm{L}}^{2} \tag{7}$$

Power to climb, considering a rate of climb RC = 2 m/s, is computed to be about 34 kW. Finally, when the cruise altitude is reached (1000 m in our case), the cruise phase could start. In order to maximize the range, the cruise speed is set to speed of maximum *L*/*D* computed as:

$$V_{\mathrm{max}\frac{L}{D}} = \sqrt{\frac{2}{\rho_{\infty}}\sqrt{\frac{k}{C_{\mathrm{D0}}}}\frac{W}{S}} \tag{8}$$

For the Rapid 200FC, a cruise speed of 141 km/h is computed with a power required for the cruise of about 17 kW, considering a propulsive efficiency of 0.75.

$$P_{\mathrm{cruise}} = \frac{D \cdot V_{\mathrm{cruise}}}{\eta_{\mathrm{P}}} \tag{9}$$

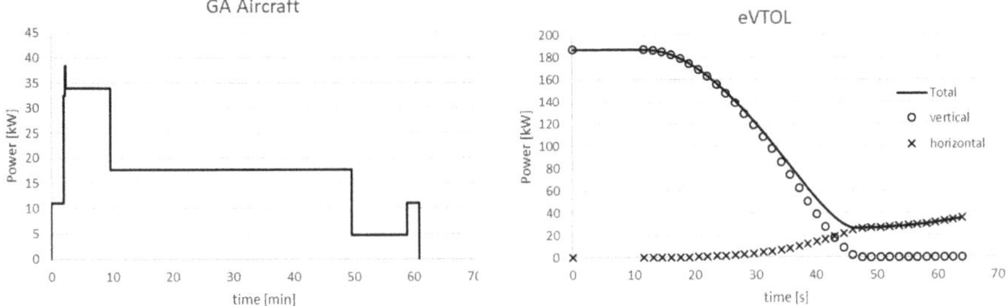


**Fig. 2** Power requirements ($P_{req}$): (**a**) GA configuration, (**b**) eVTOL configuration

In Fig. 2a, the complete power requirement considering a maximum take-off weight of 550 kg is shown.

## Electric Vertical Take-Off and Landing Powertrain

This aircraft category requires dedicated propellers or fans for take-off and landing. The system is heavy and can affect the drag of the eVTOL. The vertical thrust system can be fixed or mechanically retracted inside the fuselage during cruising flight [25, 26]. An interesting option is to use the same thrust generation system for both the hovering and the cruising phases, vectorizing the thrust as in the case adopted by Lilium [27] and implemented in this chapter.

The geometric dimensions are reported in [27]. The hover performances can be evaluated by means of disk actuator theory with some modification to account for ducted fans [26, 27]. The required power during the hovering phase is:

$$P_v = \sqrt{\frac{\left(\frac{T_v}{T_i}\right)^3}{2\rho A}} \tag{10}$$

where $T_i = 1.26$ is the thrust increment for ducted fans, $T$ is the thrust required, equal to weight, and $A$ is the disk actuator area of the entire vertical thrust system. The flight phase in which the aircraft, while accelerating, passes from hover to cruise is defined as transition. This phase is one of the most problematic, as it could lead to a loss of control power caused by moving surfaces that do not receive the necessary speed to be effective. However, despite the engineering issues and challenges, the thrust system can be suitably vectorized to produce both vertical and horizontal thrust components. Vertical thrust can be calculated simply as the difference between the weight of the aircraft and the lift produced by the wing.

$$T_v = W - L \tag{11}$$

The range of variation of the vertical thrust is between the total weight in hovering up to zero when the aircraft exceeds the stall speed, reached in this case at about 187 km/h. The horizontal thrust is calculated in order not to exceed accelerations of about 3 m/s$^2$ to have a comfortable flight experience.

The horizontal component of the power is calculated by multiplying the drag by the flight speed. These two sources of power requirements are shown in Fig. 2b.

$$P_{\mathrm{h}} = D \cdot v \tag{12}$$

From Fig. 2b, it can be seen that there is no advantage in flying at a speed slower than the stall speed providing the required vertical thrust with the motors instead of the wing. The power required would be higher, and time would be lost flying at lower speeds. To maximize the range, flight at maximum aerodynamic efficiency should be reached at cruise conditions.

Considering a quasi-sea level flight, air density $\rho = 1.225$ kg/m$^3$ and gravity acceleration $g = 9.8$ m/s$^2$ have been used. Prandtl's lifting line theory was used to compute the wing's lift and drag. Then, the additional drag of the fuselage and forward fans has been included. The drag polar is finally:

$$C_{\mathrm{D}} = 0.0163 + 0.058 \cdot {C_{\mathrm{L}}}^2 \tag{13}$$

The computed speed of maximum *L*/*D* is 244 km/h with a maximum *L*/*D* of 16.3, considering a total mass of 550 kg and a wing surface of 3.62 m$^2$.

## 2.2 Optimization Problem

The objective set for this work is to maximize the distance traveled with a given hydrogen charge of 2 kg for the GA and the eVTOL powertrains. In both cases, in addition to the hydrogen mass constraint, as the battery of the aircraft is aimed to assist the fuel cell and act as an emergency power source, it is desired that the battery state of charge (SOC) at the end of the flight is equal to that at the start. To make this possible, the fuel cell system is allowed to charge the battery during the flight when the power delivered by the fuel cell exceeds the power required by the aircraft. To allow the charge and discharge of the battery, the battery initial SOC is set as 55% and its working range is between 90% and 20%.

As stated in Eq. (1), an energy management strategy can be obtained by defining the fuel cell power profile during the flight. The mathematical model of the aircrafts presented in Sect. 2.1 was used to calculate the required electrical power profile during the flight.

To allow a comprehensive comparison of an optimized energy management strategy, we also define the base strategy as the one where the fuel cell delivers the required power, and the battery is only used when the required power is higher than the maximum fuel cell power, as stated in Eq. (14), where $\overline{P_{FC}}$ is the fuel cell power profile, $\overrightarrow{P_{\text{tot}}}$ is the required electrical power, and $P_{\text{FC, max}}$ is the maximum fuel cell power.

$$\overrightarrow{P_{FC}} = \min\left(\overrightarrow{P_{\text{tot}}}, P_{FC,\max}\right) \tag{14}$$

The optimization method used for both the aircrafts is a quasi-Newton method subject to a box constraint [28], using a forward derivative to ease the number of computations in each step.

## General Aviation Powertrain

For this powertrain, the problem to solve was: to obtain an optimal energy management strategy with a given size of the battery and fuel cell, the optimization problem for the GA powertrain is defined as:

$$\begin{aligned} &\max f\left(\overrightarrow{PFC}\right) \\ \text{subject to}\quad &H_{2\text{cons}} \leq m\text{H}_2 \\ &0 \leq PFC_j \leq PFC_{\max} \end{aligned} \tag{15}$$

where $\overrightarrow{PFC}$ is the vector of fuel cell power in the calculated points during the flight time, $f$ is the objective function as a function of the fuel cell power, $H_{2\text{cons}}$ is the $H_2$ consumed during the flight, $mH_2$ is the mass of $H_2$ loaded in the aircraft,and $PFC_{max}$ is the maximum power of the fuel cell.

The vector of fuel cell power can have any desired length, that is, we can calculate the fuel cell power with any desired time precision. However, the greater the size of the vector, the more computationally expensive is the problem. For this work, every segment, i.e.,taxi, ground roll, climb, cruise, etc., was divided into seven points equally spaced in time where the fuel cell power is defined.

For this powertrain, the fuel cell maximum power is defined as 25 kW and a battery of 6.6 kWh, in line with the power sources used in the ENFICA project [29].

As stated before, the SOC difference between the start and the end of the flight is required to be the same. This requirement can be relaxed by stating it as a penalization term in the objective function, a term proportional to the absolute value of the SOC difference. In addition, the battery power profile is calculated for the entire flight, and the maximum required current is calculated. If the maximum current exceeds the value determined by the battery manufactured, a proportional penalization term is added to ensure that the battery size is appropriate. The objective function for the GA powertrain is stated as follows:

$$f\left(\left\{\overrightarrow{PFC},N_{\mathrm{BAT}}\right\}\right)=\rho_0\cdot t\left(\left\{\overrightarrow{PFC},N_{\mathrm{BAT}}\right\}\right)-\rho_1\cdot\left|SOC_{t=\mathrm{end}}-SOC_{t=0}\right| \tag{16}$$

where $t\left(\left\{\overrightarrow{PFC},N_{\mathrm{BAT}}\right\}\right)$ is the flight time as a function of the fuel cell power vector and the number of batteries, SOC is the state of charge of the battery, and $\rho_0 > 0$ and $\rho_1 > 0$ are normalization coefficients properly chosen. In this case, $\rho_0 = 0.0971$ and $\rho_1 = 5000$.

The fight time and the SOC for objective functions are calculated as follows. First, for a fuel cell power profile given by the optimization algorithm, the consumption of all the flight segments except cruise is calculated. The cruise segment duration is calculated based on the mass of the remaining hydrogen. Once the duration of the flight is set, the fuel cell model is run with the entire flight power profile and the hydrogen consumption constraint is checked, if it is not met, the duration of the cruise segment is adjusted accordingly, and the time of the flight is given. Once the flight time is defined, the battery power profile is simulated with the battery model, giving the current ($\vec{I}$) and SOC profiles.

## Electric Vertical Take-Off and Landing Powertrain

For this powertrain, in order to add further complexity to the problem, three optimizations are proposed. For all the optimizations, the objective was the same, i.e., maximize the flight range, while trying to minimize the SOC difference. First, given an adequate fuel cell size and using the base strategy stated in Eq. (14), the battery size is optimized (Eq. 17). The second optimization consists of optimizing the size of both the power sources using the base strategy (Eq. 18). Finally, the third optimization gives an optimal energy management and the optimal sizing of the fuel cell and the battery (Eq. 19).

$$\begin{gathered}\max f\left(N_{\mathrm{BAT}}\right)\\ \text{subject to}\quad H_{2\mathrm{cons}}\le m\mathrm{H}_2\\ f\left(N_{\mathrm{BAT}}\right)=\rho_0\cdot t\left(N_{\mathrm{BAT}}\right)-\rho_1\cdot\left|SOC_{t=\mathrm{end}}-SOC_{t=0}\right|-\rho_2\cdot MCP\\ MCP=\begin{cases}\dfrac{\max\left(\vec{I}\right)}{I_{\max}} & \text{if }\max\left(\vec{I}\right)>I_{\max}\\ 0 & \text{if }\max\left(\vec{I}\right)\le I_{\max}\end{cases}\end{gathered} \tag{17}$$

$$\max f\left(\left\{P_{FC,\max}, N_{\text{BAT}}\right\}\right)$$

$$\text{subject to} \quad H_{2\text{cons}} \le m\text{H}_2$$

$$f\left(\left\{\overrightarrow{PFC}, N_{\text{BAT}}\right\}\right) = \rho_0 \cdot t\left(\left\{\overrightarrow{PFC}, N_{\text{BAT}}\right\}\right) - \rho_1 \cdot \left|SOC_{t=\text{end}} - SOC_{t=0}\right| - \rho_2 \cdot MCP \tag{18}$$

$$\max f\left(\left\{\widehat{\overrightarrow{P_{FC}}}, P_{FC,\max}, N_{\text{BAT}}\right\}\right)$$

$$\text{subject to} \quad \begin{array}{l} H_{2\text{cons}} \le m\text{H}_2 \\ 0 \le \widehat{P_{FCi}} \le 1 \end{array} \tag{19}$$

$$f\left(\left\{\widehat{\overrightarrow{P_{FC}}}, P_{FC,\max}, N_{\text{BAT}}\right\}\right) = \rho_0 \cdot t\left(\left\{\widehat{\overrightarrow{P_{FC}}}, P_{FC,\max}, N_{\text{BAT}}\right\}\right) - \rho_1 \cdot \left|SOC_{t=\text{end}} - SOC_{t=0}\right| - \rho_2 \cdot MCP$$

where $\widehat{P_{FCi}} \in [0,1]$ are the elements of the normalized fuel cell power profile, and the fuel cell power profile is expressed as:

$$\overrightarrow{P_{FC}} = P_{FC,\max} \widehat{\overrightarrow{P_{FC}}} \tag{20}$$

From Eqs. (17) to (19), MCP im, $\max\left(\vec{I}\right)$ is the maximum current delivered by the battery, and $I_{\max}$ is the battery maximum discharge current given by the manufacturer. As stated before, $\rho_0 > 0$, $\rho_1 > 0$, and $\rho_2 > 0$ are normalization coefficients properly chosen. In all the problems, $\rho_0 = 0.2375$, $\rho_1 = 5000$, and $\rho_2 = 5000$.

For the eVTOL powertrain objective function, the take-off and landing profiles are calculated first, and then the cruise duration is calculated based on the remaining hydrogen mass. Then the complete trip is simulated with the fuel cell and the battery model to ensure that the constraints are met, and correct the cruise time in case they are not.

## 2.3 *Optimization Results*

For both, the GA and the eVTOL powertrains, the results of the optimization show marginal improvements in terms of range due to the constraints of the problem.

In Table 1, the simulation results of the base strategy and the optimized strategy for the GA powertrain are shown. Besides the flight time and traveled distance, it is interesting to measure the energy consumption. When two or more power sources are used, an equivalent energy can be calculated. One of the approaches used to define the equivalent energy consumption is the one stated in Eq. (21), where $E_{\text{eq}}$ is the equivalent energy, $m\text{H}_2$ is the hydrogen mass consumed in the flight (2 kg), $LHV_{\text{H}_2} = 120\ MJ/\text{kg}$ is the lower heating value of hydrogen and $E_{\text{bat}}$ is the battery energy or size.

**Table 1** Optimization results for the GA powertrain

| | Base strategy | Optimized strategy |
|---|---|---|
| Flight time (h) | 2.86 | 2.94 |
| Cruise speed (m/s) | 39.3 | |
| Distance covered (km) | 345.61 | 355.23 |
| ΔSOC (%) | −0.33 | −1 |
| Objective function value | 992.65 | 1003.06 |
| FC max. power (kW) | 25 | 25 |
| Battery size (kWh) | 6.6 | 6.6 |
| Equivalent energy (kWh) | 66.69 | 66.73 |
| Eq. energy consumption (MJ/km) | 0.6947 | 0.6762 |
| Aircraft weight (kg) | 479 | |

$$E_{\mathrm{eq}} = m\mathrm{H}_2 \cdot LHV_{\mathrm{H}_2} + \left(SOC_{t=0} - SOC_{t=\mathrm{end}}\right) \cdot E_{\mathrm{bat}} \tag{21}$$

The optimized strategy for the GA powertrain shows a 2.78% improvement in the distance traveled from 345 to 355 km. ΔSOC refers to SOC difference between the start and the end of the flight, and as stated in Sect. 2.2.1, ΔSOC = 0 constraint was relaxed as a penalization term, which allows solutions where ΔSOC ≠ 0. For the GA powertrain optimization, the SOC difference between the start and end of the flight varies from 0.33% with the base strategy to 1% with the optimized strategy. The 0.77% difference between the strategies amounts to 0.05 kWh of battery energy. During the cruise, this energy corresponds to an extra 12 s or 0.49 km, showing that the improvement for the optimized strategy is not only due to higher battery use, i.e., when the optimized strategy is used, the aircraft is able to travel an extra 10 km.

The effect of the optimization can be seen in the value of the objective function, which increases its value by 1%.

While the equivalent energy increases in the optimized strategy due to a higher ΔSOC, the equivalent energy consumption per kilometer shows a 2.7% decrease in favor of the optimized strategy.

Figures 3 and 4 show the power profiles of the electrical power required and each of the power sources in the left axis and the SOC value in the right axis during the flight for the base strategy and the optimized strategy, respectively. In Fig. 3, it can be seen that for the base strategy, the battery is only used during the take-off and climbing stages.

In Fig. 4, it can be seen that the optimized strategy uses the start and end of the cruise, and the descent stages to charge the battery, using it during take-off, climbing, and cruise stages allowing the cell to work at a lower power.

The results of the different optimization for the eVTOL powertrain can be seen in Table 2. As stated before, three optimizations were performed: a battery sizing with the base strategy (referred later as base strategy), a sizing of the battery and the fuel cell power using the base strategy (referred later as base strategy with optimal

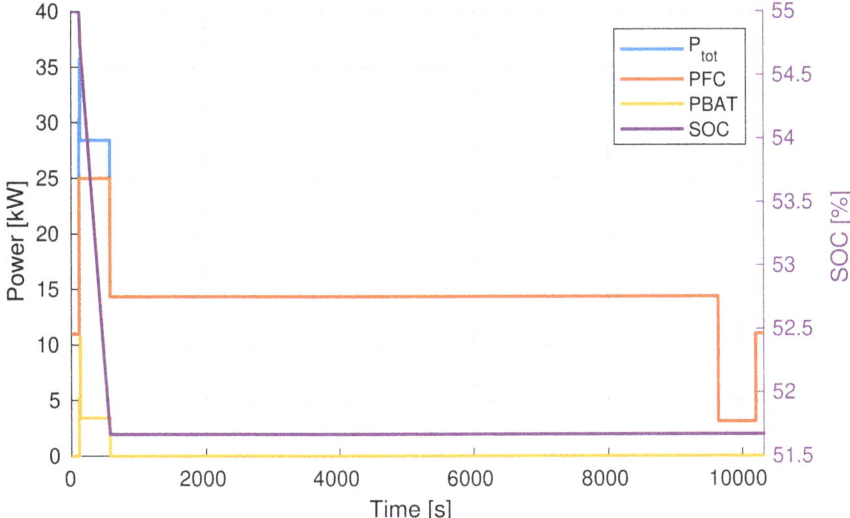


**Fig. 3** Power profiles for the base case

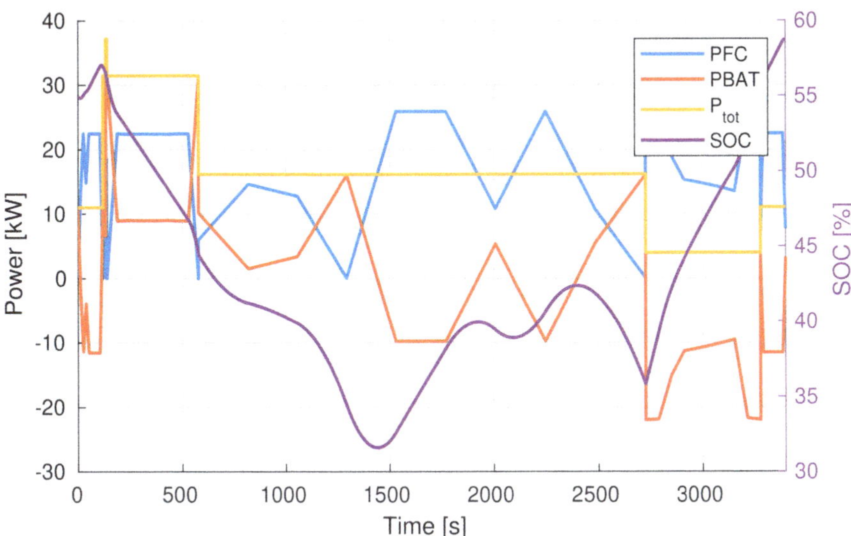


**Fig. 4** Power profiles obtained after the energy management strategy optimization

sizing), and an optimization of the battery and the fuel cell size along with the energy management (optimized strategy).

Comparing the base strategy with the optimized one, a decrease of 7.14 km for the optimized strategy can be seen, nonetheless, the optimized strategy allows a zero ΔSOC difference, highlighting how the proposed optimization strategy can be adopted to meet the minimum SOC difference requirements of an effective

**Table 2** Optimization results for the eVTOL powertrain

| | Base strategy | Base strategy with optimized sizing | Optimized strategy |
|---|---|---|---|
| Flight time (h) | 1.13 | 1.21 | 1.17 |
| Cruise speed (m/s) | 67.7 | | |
| Distance covered (km) | 277.80 | 289.10 | 270.66 |
| ΔSOC (%) | −33.99 | −30.02 | 0 |
| Objective function value | −729.9 | −463.5 | 1000 |
| FC max. power (kW) | 40 | 29.88 | 30.97 |
| Battery size (kWh) | 13.32 | 14.27 | 9.18 |
| Equivalent energy (kWh) | 71.19 | 70.95 | 66.67 |
| Energy consumption (MJ/km) | 0.9225 | 0.8835 | 0.8867 |
| Aircraft weight (kg) | 583 | 549 | 519 |

on-demand taxi service operations that require a frequent recharge of eVTOLs within a limited time. In particular, the optimized solution presents an aircraft that is able to guarantee a distance traveled comparable to the base strategy one but with zero SOC difference thanks to an optimal sizing of the battery and fuel cell system dimensions, leading to a lighter and therefore more performing aircraft (optimized strategy).

If the most stringent requirement is to obtain a high distance traveled while the objective function penalizes the ΔSOC requirement, when the optimizations are performed on the sizing of both power sources (base strategy with optimized sizing), a similar solution to the base strategy is found in terms of ΔSOC but with a lighter aircraft with a range that has increased by 11.3 km.

If the aircraft using the optimized strategy were to use 30.02% of the battery, to match the battery consumption of the base strategy with optimized sizing, that would amount to 2.76 kWh that during the cruise stage could be used to travel 361 extra seconds covering a distance of 24.43 km, when the range difference between strategies is only 18.44 km. This result shows that the optimized strategy benefits do not come from the ΔSOC difference.

In Figs. 5, 6, 7, 8, 9, and 10, the power profiles for the eVTOL aircraft, fuel cell, and battery are shown alongside the SOC during the flight.

Figures 5 and 7 show the power profiles for the base strategy and optimal battery sizing. Figures 6 and 8 show a closeup of the start and the end of the flight, where the eVTOL powertrain power requirements concentrate.

Figures 5 and 6 show the base strategy evolution during the flight; it can be seen that the maximum fuel cell power of 40 kW is slightly higher than the required power in the cruise stage, while the maximum power required by the aircraft is five times higher. This elevated power requirement is met by a surge in the battery power.

In Figs. 7 and 8, the base strategy for the fuel cell and battery optimal sizing can be seen. This optimal sizing decreases the fuel cell size from 40 to 29.88 kW to meet the cruise-required power and increases the battery size from 13.32 to 14.27 kWh to

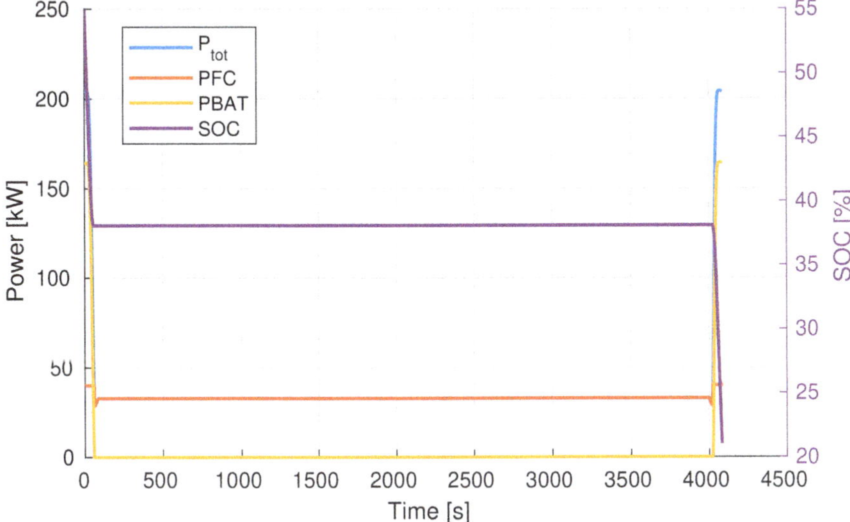


**Fig. 5** Power profiles obtained for the base strategy with battery size optimization

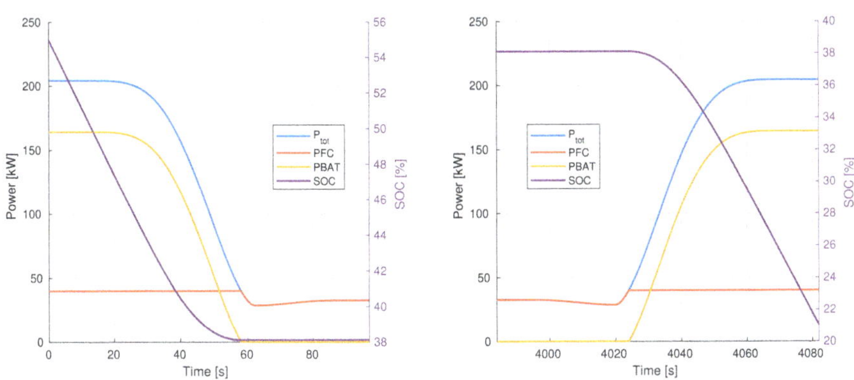


**Fig. 6** Take-off and landing power profiles for the base strategy with battery size optimization

be able to contribute during the take-off and landing stages, having a lower ΔSOC at the end of the flight.

Since the base strategy (Eq. 14) does not contemplate battery charging, in the case of the battery sizing and the fuel cell and battery sizing (from Figs. 5 to 8), it can be seen that the battery is only discharged, especially during the take-off and landing.

Lastly, Figs. 9 and 10 show the power profiles of the optimized EMS with optimal sizing.

It can be seen that in this optimal EMS, the fuel cell operates at maximum power during the whole flight. During the cruise stage, the fuel cell power is higher than the aircraft-required power allowing charging of the battery. With this strategy, the

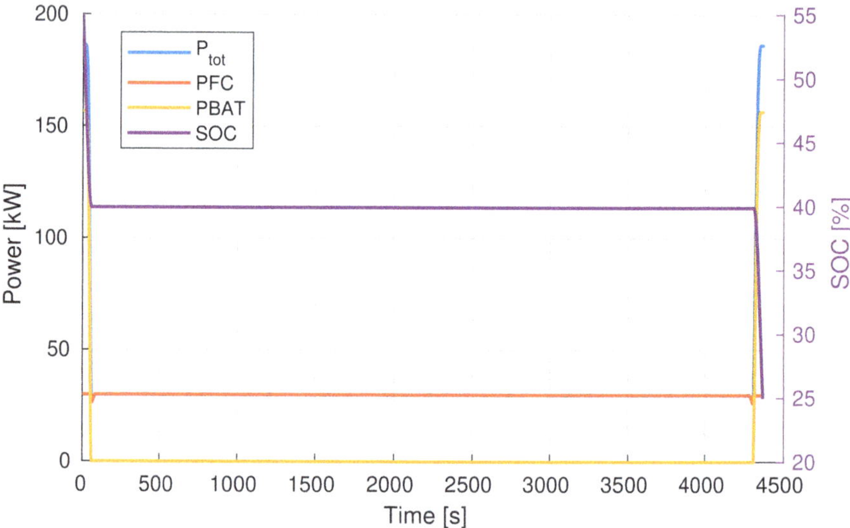


**Fig. 7** Power profiles obtained for the base strategy with battery and fuel cell size optimization

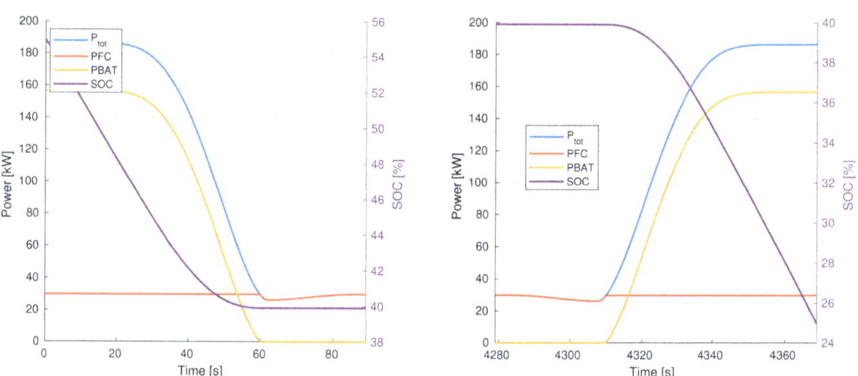


**Fig. 8** Take-off and landing power profiles for the base strategy with battery and fuel cell size optimization

aircraft is able to end the flight with a ΔSOC of zero after the final battery discharge during the landing.

## 3 Concluding Remarks

In this work, an optimization method is proposed to obtain two optimized EMS to be applied to two different types of all-electric aircraft. The studied aircraft is designed to operate as general aviation on-demand taxi service using the same

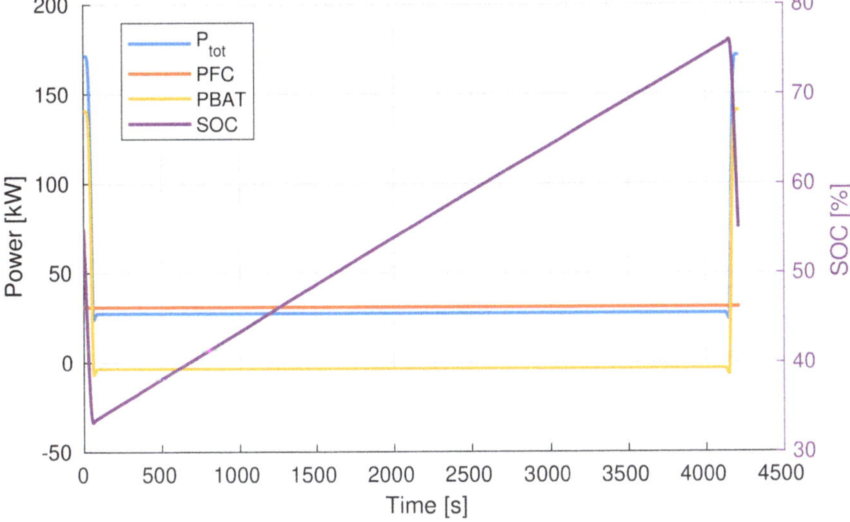


**Fig. 9** Power profiles obtained for the optimized size and energy management strategy

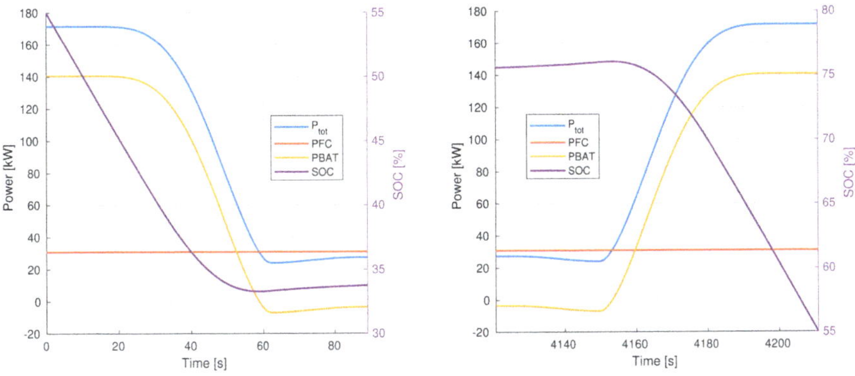


**Fig. 10** Take-off and landing power profiles for the optimized size and energy management strategy

powertrain configuration: hybrid aircraft powered by a hydrogen-fueled fuel cell and a battery pack. The battery of the aircraft is not meant to be charged at the end of the flight in order to maximize the operational time of the aircraft by avoiding the battery charge.

The multiplicity of power sources of these aircraft increases the complexity of the sizing of power sources and adds the problem of energy management.

To maximize the traveled distance while keeping the SOC difference at a minimum and observing the power sources restriction, the optimization of the energy management was performed. In addition, for the eVTOL powertrain, the optimal power source sizing was performed for a base strategy and for the optimized strategy the sizing was performed alongside the energy management optimization. The simulation of the models of the aircraft and its power sources enables the optimization of the power sources sizing and yields an optimal EMS, while considering important aspects as the maximum battery current, working temperature, and SOC, among others.

The analysis of the results show that using the proposed method, improved EMSs can be obtained. For the GA powertrain, the optimized EMS improves the traveled distance by 2.78% and reduces the equivalent energy consumption by 2.73%. In the eVTOL powertrain, a small decrease in the traveled distance is observed in comparison with other strategies, nonetheless, the optimized EMS guarantees that the battery start and end the flight with the same SOC allowing a seamless operation without the need to stop a prolonged time to charge the battery, or avoids to shorten the battery lifetime if a fast charge is to be implemented.

## References

A. García-Olivares, J. Solé, O. Osychenko, Transportation in a 100% renewable energy system, Energy Convers. Manag. 158 (2018) 266–285. https://doi.org/10.1016/J.ENCONMAN.2017.12.053.

United Nations, Emissions Gap Emissions Gap Report 2020, 2020. https://www.unenvironment.org/interactive/emissions-gap-report/2019/.

European Commission, Reducing emissions from aviation, Eur. Comm. (2018). https://ec.europa.eu/clima/eu-action/transport-emissions/reducing-emissions-aviation_en (accessed December 10, 2021).

Airbus, Airbus studies fuel cell pods for future aircraft, Fuel Cells Bull. 2021 (2021) 6. https://doi.org/10.1016/S1464-2859(21)00018-3.

C.E.D. Riboldi, Energy-optimal off-design power management for hybrid-electric aircraft, Aerosp. Sci. Technol. 95 (2019) 105507. https://doi.org/10.1016/j.ast.2019.105507.

C.E.D. Riboldi, An optimal approach to the preliminary design of small hybrid-electric aircraft, Aerosp. Sci. Technol. 81 (2018) 14–31. https://doi.org/10.1016/j.ast.2018.07.042.

G. Romeo, F. Borello, G. Correa, E. Cestino, ENFICA-FC: Design of transport aircraft powered by fuel cell; flight test of zero emission 2-seater aircraft powered by fuel cells fueled by hydrogen, Int. J. Hydrogen Energy. 38 (2013) 469–479. https://doi.org/10.1016/j.ijhydene.2012.09.064.

D.P. Bertsekas, Nonlinear Programming, 3rd Editio, Athena Scientific, 2016. http://www.athenasc.com/nonlinbook.html.

S.G. Johnson, The NLopt nonlinear-optimization package, (2008). https://github.com/stevengj/nlopt.

P. Virtanen, R. Gommers, T.E. Oliphant, M. Haberland, T. Reddy, D. Cournapeau, E. Burovski, P. Peterson, W. Weckesser, J. Bright, S.J. van der Walt, M. Brett, J. Wilson, K.J. Millman, N. Mayorov, A.R.J. Nelson, E. Jones, R. Kern, E. Larson, C.J. Carey, Ilhan Polat, Y. Feng, E.W. Moore, J. VanderPlas, D. Laxalde, J. Perktold, R. Cimrman, I. Henriksen, E.A. Quintero, C.R. Harris, A.M. Archibald, A.H. Ribeiro, F. Pedregosa, P. van Mulbregt, SciPy 1.0 Contributors, SciPy 1.0: Fundamental Algorithms for Scientific Computing in Python, Nat. Methods. 17 (2020) 261–272. https://doi.org/10.1038/s41592-019-0686-2.

H. Rezk, A.M. Nassef, M.A. Abdelkareem, A.H. Alami, A. Fathy, Comparison among various energy management strategies for reducing hydrogen consumption in a hybrid fuel cell/supercapacitor/battery system, Int. J. Hydrogen Energy. 46 (2021) 6110–6126. https://doi.org/10.1016/j.ijhydene.2019.11.195.

I.-S. Sorlei, N. Bizon, P. Thounthong, M. Varlam, E. Carcadea, M. Culcer, M. Iliescu, M. Raceanu, Fuel Cell Electric Vehicles—A Brief Review of Current Topologies and Energy Management Strategies, Energies. 14 (2021) 252. https://doi.org/10.3390/en14010252.

T. Teng, X. Zhang, H. Dong, Q. Xue, A comprehensive review of energy management optimization strategies for fuel cell passenger vehicle, Int. J. Hydrogen Energy. (2020). https://doi.org/10.1016/J.IJHYDENE.2019.12.202.

N. Bizon, Optimization Algorithms and Energy Management Strategies, in: 2020: pp. 57–105. https://doi.org/10.1007/978-3-030-40241-9_3.

N. Bizon, Real-time optimization strategies of Fuel Cell Hybrid Power Systems based on Load-following control: A new strategy, and a comparative study of topologies and fuel economy obtained, Appl. Energy. 241 (2019) 444–460. https://doi.org/10.1016/j.apenergy.2019.03.026.

Y. Zhou, A. Ravey, M.-C. Péra, A survey on driving prediction techniques for predictive energy management of plug-in hybrid electric vehicles, J. Power Sources. 412 (2019) 480–495. https://doi.org/10.1016/j.jpowsour.2018.11.085.

M. Doff-Sotta, M. Cannon, M. Bacic, Optimal energy management for hybrid electric aircraft, IFAC-PapersOnLine. 53 (2020) 6043–6049. https://doi.org/10.1016/j.ifacol.2020.12.1672.

J. Zhao, H.S. Ramadan, M. Becherif, Metaheuristic-based energy management strategies for fuel cell emergency power unit in electrical aircraft, Int. J. Hydrogen Energy. 44 (2019) 2390–2406. https://doi.org/10.1016/j.ijhydene.2018.07.131.

S. Li, C. Gu, M. Xu, J. Li, P. Zhao, S. Cheng, Optimal power system design and energy management for more electric aircrafts, J. Power Sources. 512 (2021) 230473. https://doi.org/10.1016/j.jpowsour.2021.230473.

G. Correa, F. Borello, M. Santarelli, Sensitivity analysis of stack power uncertainty in a PEMFC-based powertrain for aircraft application, Int. J. Hydrogen Energy. 40 (2015) 10354–10365. https://doi.org/10.1016/J.IJHYDENE.2015.05.133.

G. Correa, M. Santarelli, F. Borello, E. Cestino, G. Romeo, Flight test validation of the dynamic model of a fuel cell system for ultra-light aircraft, Proc. Inst. Mech. Eng. Part G J. Aerosp. Eng. 229 (2015) 917–932. https://doi.org/10.1177/0954410014541081.

G. Romeo, G. Correa, F. Borello, E. Cestino, M. Santarelli, Air Cooling of a Two-Seater Fuel Cell–Powered Aircraft: Dynamic Modeling and Comparison with Experimental Data, J. Aerosp. Eng. 25 (2012) 356–368. https://doi.org/10.1061/(ASCE)AS.1943-5525.0000138.

G. Correa, P. Muñoz, T. Falaguerra, C.R. Rodriguez, Performance comparison of conventional, hybrid, hydrogen and electric urban buses using well to wheel analysis, Energy. 141 (2017) 537–549. https://doi.org/10.1016/j.energy.2017.09.066.

D. Raymer, Aircraft Design: A Conceptual Approach, Sixth Edition, American Institute of Aeronautics and Astronautics, Inc., Washington, DC, 2018. https://doi.org/10.2514/4.104909.

A. Bacchini, E. Cestino, B. Van Magill, D. Verstraete, Impact of lift propeller drag on the performance of eVTOL lift+cruise aircraft, Aerosp. Sci. Technol. 109 (2021) 106429. https://doi.org/10.1016/J.AST.2020.106429.

A. Bacchini, E. Cestino, Key aspects of electric vertical take-off and landing conceptual design, Proc. Inst. Mech. Eng. Part G J. Aerosp. Eng. 234 (2020) 774–787. https://doi.org/10.1177/0954410019884174.

A. Bacchini, E. Cestino, Electric VTOL Configurations Comparison, Aerospace. 6 (2019) 26. https://doi.org/10.3390/aerospace6030026

P.M. Muñoz, G. Correa, M.E. Gaudiano, D. Fernández, Energy management control design for fuel cell hybrid electric vehicles using neural networks, Int. J. Hydrogen Energy. 42 (2017) 28932–28944. https://doi.org/10.1016/j.ijhydene.2017.09.169.

G. Romeo, F. Borello, G. Correa, Set-up and test flights of an all-electric 2-seater aeroplane powered by fuel cells, J. Aircr. 49 (2011).

# Hydrogen Infrastructure and Logistics in Airports

**Maršenka Marksel, Rok Kamnik, Stanislav Božičnik, and Anita Prapotnik Brdnik**

## 1 Introduction

Although aviation is not currently one of the leading causes of global warming, recent trends indicate that it could become a significant factor in the coming decades. Today, aviation is responsible for about 2% of global and 3% of total EU greenhouse gas emissions. Without action to reduce emissions, global aviation emissions could increase by over 300% by 2050 [1]. Various pathways and scenarios for minimazing the environmental impact of aviation in the coming years [2–4] conclude that in the short term, various measures such as operational and infrastructure improvements, economic measures, and evolutionary technologies in aircraft and configurations are most likely to reduce emissions. They add that radical improvements, and technologies, among others also hydrogen aircraft, will predominate in efforts to reduce emissions in the long term.

Currently, hydrogen is used to primarily in the chemical industry and to a lesser extent in other industries. It is used in a wide range of transportation such as cars, buses, trains, ferries, etc., and there is also great potential for its use in aviation [5]. Compared to kerosene, the energy density of hydrogen is 2.75 times higher than the energy density of kerosene (33 kWh/kg compared to 12 kWh/kg). As weight plays a crucial role in aviation, its high energy density makes it an interesting fuel for aviation, which also motivated the first research steps in this direction. Unfortunately, the critical drawback of hydrogen is its gaseous state at ambient temperature and its low volume density. Therefore, it must be kept in cryogenic tanks below −253 °C in

M. Marksel (✉) · R. Kamnik · S. Božičnik · A. P. Brdnik
Faculty of Civil Engineering, Transportation Engineering and Architecture, University of Maribor, Maribor, Slovenia
e-mail: marsenka.marksel@um.si; rok.kamnik@um.si; stane.bozicnik@um.si; anita.prapotnik@um.si

 
C. O. Colpan, A. Kovač (eds.), *Fuel Cell and Hydrogen Technologies in Aviation*, Sustainable Aviation, https://doi.org/10.1007/978-3-030-99018-3_6

a liquid form or in pressurised tanks, usually at 350 or 700 bar. Since these tanks are both large and heavy, any advantages that result from the high energy density can be lost due to high weight of the tanks. Another aspect of hydrogen, which needs to be considered due to its physical and chemical properties, is safety. Although safety issues were considered a significant disadvantage in the past, the experience and technological advances gained have made hydrogen operations manageable today. Despite some safety procedures that need to be considered when using hydrogen, it has apparent advantages as aviation fuel in emitting fewer toxic and greenhouse gases compared to conventional jet fuel. When hydrogen is burned, no $CO_2$ and $SO_x$ emission are produced, while $NO_x$ emissions are still present. If hydrogen is used in fuel cells, the only by-product is water vapour [5].

Nevertheless, if hydrogen is considered as a future green fuel, its production must be environmentally sound; currently, 95% of hydrogen is produced by steam reforming, mostly from methane. This must change in the future, and production must shift towards electrolysis using electricity from renewable sources, like wind and solar energy. Another ecologically sound solution for aircraft propulsion would be synthetic fuels, obtained by biding carbon dioxide from the air using energy obtained from renewable sources. Compared to hydrogen, production of such alternative fuels would require more energy, leading to higher costs [6]. On the other hand, synthetic fuels do not require cryogenic or pressurized tanks and are technologically easier to implement. An all electric propulsion, system powered by batteries or solar cells is an attractive green alternative solution to the current aircraft propulsion system. Due to deficient specific energy, batteries and solar cells are only considered as energy sources for small aircraft types (mainly for general aviation and partially for regional aircraft) [7]. When it comes to an environmentally friendly propulsion technology that is also suitable for commercial applications, the hydrogen-powered aircraft represents a viable solution.

The development of hydrogen logistics in aviation and hydrogen infrastructure at the airport will be closely related to the technical characteristics and the development of hydrogen aircraft. Therefore, in the following introductory subsections, we will shortly describe hydrogen aircraft, their basic characteristics, their history, and their future development perspectives. This should give the reader an idea of the pace of development and the expected demands of type and quantity of hydrogen. Section 2 describes hydrogen logistics—from production to transportation and storage. Section 3 will concentrate on infrastructure needs on large and small airports, refuelling procedure and hydrogen supply for handling ground vehicles. Section 4 will be devoted to hydrogen safety.

### *1.1 Hydrogen Aircraft*

A hydrogen-powered aircraft is an aircraft that uses hydrogen as an energy source. Hydrogen can either be burned in a jet engine or other internal combustion engine (e.g. aircraft powered by hydrogen (H2) turbines) or used in a fuel cell to generate electricity to drive a propeller (e.g. fuel cell aircraft). Both liquid hydrogen and compressed hydrogen can be used as a fuel.

**Aircraft powered by H2 turbines** are like conventional aircraft with internal combustion engine (ICE), expect that, hydrogen is burned in the H2 turbines instead of kerosene. The hydrogen engines will remain similar in their architecture to existing jet engines, but with some adoptions, such as fuel pumps and control units, combustion chambers, and an additional heat exchanger to vaporise liquid hydrogen (LH2). Aircraft design will have to change due to the more spacious fuel reservoirs [8]. According to some testing, such hydrogen engines are expected to have, about 64% less Specific Fuel Consumption (SFC) than conventional jet engines. Furthermore, these engines are expected to be 1–5% more efficient in generating thrust from the given energy content. They emit no $CO_2$ and emit about 5–25% of the $NO_x$ gases of their conventional counterparts [9]. Long-range aircraft are expected to be up to 12% more energy efficient than conventional kerosene aircraft because a reduction in fuel weight mitigates the increase of the aircraft weight due to large and heavy cryogenic tanks. For short-range aircraft, the ratio between fuel weight benefit and aircraft weight penalty is not so favourable, leading to up to 18% worse energy efficiency than conventional kerosene aircraft [10]. Theoretically, the first hydrogen cryogenic aircrafts were to be developed by 2020 and commercially deployed by 2040 (estimation) [8, 11].

**Fuel cell aircraft** use hydrogen in fuel cells to produce electrical energy and water. Among the different fuel cells, the Proton Exchange Membrane Fuel Cell (PEMFC) is most suitable for aircraft applications [12]. The electrical energy produced by the fuel cells drives the electrical generator that drives the propeller. If needed, fuel cells can be combined with batteries to generate additional power during take-off and climb. Because the fuel cell aircraft is propeller-driven, it is therefore similar to the current piston and turboprop aircraft. This type of aircraft is best suited for low-altitude and short-haul flights. Compared to internal combustion engines, fuel cells are silent, generate a slight vibration, and produce zero $NO_x$ emissions, making them very attractive for many applications. They also reach higher efficiency (between 40 and 60%) than combustion engines (approximately 30%). Nevertheless, they produce a lot of water vapour, which is also a greenhouse gas and acts as an air pollutant at high altitudes. The main challenge with fuel cell aircraft is the low power-to-weight ratio of fuel cells, leading to even heavier aircraft designs than hydrogen-ICE technology. However, fuel cell technology is constantly evolving so that the power-to-weight ratio of fuel cells is increasing every year [13]. It is expected that fuel cell technology will be initially introduced in the small regional aircraft segment with up to 80 passenger seats.

## 1.2 Hydrogen Aircraft History

The early attempts to use liquid hydrogen (LH2) as a fuel in aircraft were in the military [14]. A modified Wright J-65 turbojet engine was installed in a US Air Force B-57 twin-engine bomber that performed its maiden flight in February 1957.

The aircraft could fly for 20 min on liquid hydrogen pressurized with helium in one of its two engines [15]. After the first successful flight, the potential of hydrogen as an aircraft fuel resulted in the development of a high speed and high-altitude military aircraft project, CL-400. Unfortunately, this project was cancelled due to the high cost, difficulty in achieving the desired range, and doubts about saftey of liquid hydrogen. Later, development shifted to using liquid hydrogen as rocket fuel for space missions. It was successfully used in the Apollo and Space Shuttle programmes and thus became the primary propellant in space exploration programmes. At that time, the use of hydrogen as fuel in commercial aviation had no obvious advantages. As a result, there were few attempts in this direction until the oil crisis, and environmental awakening triggered new research interest. The 1980s were marked with the Soviets testing hydrogen on a modified TU-154 aircraft (renamed TU-155), the first experimental civil aircraft running on liquid hydrogen. Research continued in the 1990s with collaboration between the Soviet Union and Germany on liquid-hydrogen-fueled commercial prototypes similar to the A310 with an estimated range of 500 miles [15].

In today's environmental crisis, it was recognised that hydrogen has good potential as a clean energy source in aviation, leding to new research opportunities to develop a commercial hydrogen aircraft. In 2000, the European Commission funded the CRYOPLANE project to analyse aircraft fuelled by LH2 [16]. Although burning hydrogen does not emit $CO_2$, it still emits $NO_X$ gases. The increase of emissions into the environment has led to several European Commissions (EC) initiatives to developt of fuel cell technology, powered by either gaseous or liquid hydrogen that emits water vapour. In 2005, AeroVironment tested a miniature version of the global observer prototype called "Odyssey" powered solely by fuel cells, followed by the first flight of the full-sized fuel cell powered Global Observer in 2011. The aircraft went down in history as the world's first LH2-powered Unmanned Aerial Vehicle (UAV) that managed to complete more than one hour of flight using only liquefied hydrogen [17]. Problems occurred during the test flights, resulting in an incident and a delay of the second aircraft production. Since then, the AeroVironment has been engaged in intense competition with its rival, Boeing Phantom Works, which successfully flew its Boeing Fuel Cell Demonstrator—the first crewed aircraft using fuel cell technology and batteries—in 2008 [18] and the Phantom Eye—a long range drone powered solely by liquid hydrogen in 2012 [19].

The first manned aircraft using solely fuel cell technology was **DLR-H2** (see Fig. 1a), developed by German Aerospace Centre (DLR) in 2009 [20], while the four-seater fuel cell aircraft **DLR-HY4** (now further developed by the company H2FLY GmbH which made its maiden flight in September 2016) is the first aircraft powered by fuel cell technology using compressed hydrogen that can be used for commercial passenger mobility [21] (see Fig. 1b).

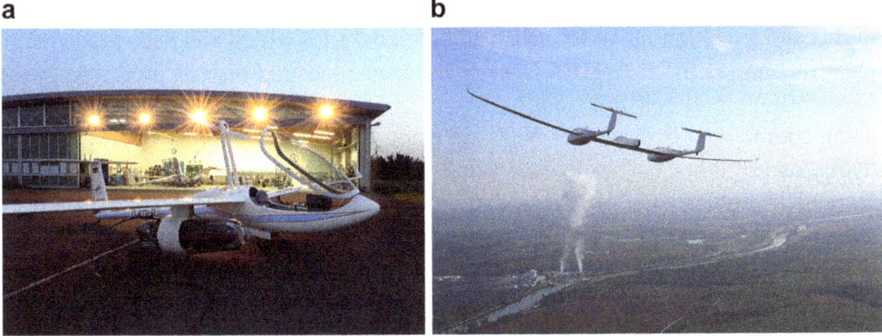

**Fig. 1** (**a**) Antares DLR-H2 (Source: DLR (CC-BY-NC-ND 3.0) [20]). (**b**) H2FLY HY4 (Source: Courtesy of H2FLY, 2021)

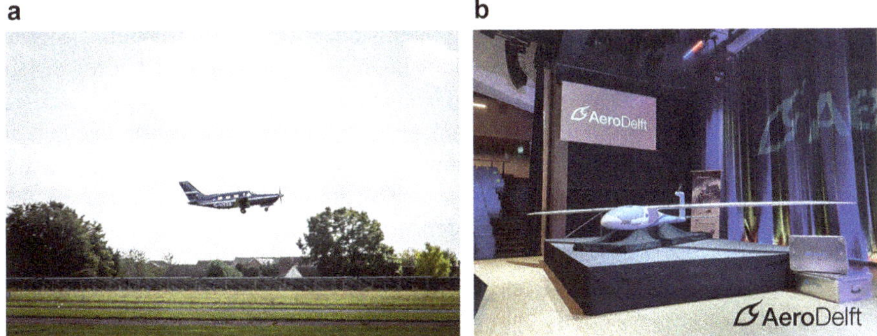


**Fig. 2** (**a**) ZeroAvia HyFlyer (Source: Courtesy of ZeroAvia, 2021). (**b**) AeroDelft Phoenix PT (Source: Courtesy of AeroDelft, 2021)

## *1.3 Further Expected Applications of Fuel Cell Aircraft*

In the recent development of hydrogen fuel cells, several promising research- and product-oriented projects have been identified. **ZeroAvia's HyFlyer I** project (see Fig. 2a) utilized a Piper M-class six-seater aircraft with a hydrogen-electric propulsion and made the world's first hydrogen-electric fuel cell-powered flight in a commercial-grade aircraft in September 2020. The company is expected to start supplying hydrogen-electric powertrain technology to aircraft manufacturers and operators in 2024. **AeroDelft** plans to fly the **Phoenix PT** (see Fig. 2b) in 2022, which will be the world's first fuel cell aircraft powered by liquid hydrogen [22]. Extensive fuel cell research continues with **NASA** developing a cryogenic hydrogen fuel cell system for powering all-electric aircraft within its Center for Cryogenic High-Efficiency Electrical Technologies for Aircraft (2019–2021). **Alaka'i**

**Technologies** is developing a liquid hydrogen fuel cell air-taxi capable of four-hour flights and a 643 km range for urban and regional trips which had been predicted to enter the market in 2020. **HES Energy Systems** plans to finish its prototype of the Element One fuel cell four-seat passenger aircraft in 2025. It will fly for 500 km using compressed hydrogen or 5000 km using liquid hydrogen [23]. Most fuel cell hydrogen prototypes are built on a small-scale level, up to 4–10 passengers, while theoretical studies exist for a hypothetical fuel cell aircraft capable of transporting 140–180 passengers [24]. Despite great progress in hydrogen-powered aviation, the realization of such aircraft is not expected before 2030–2035 [6] or even before 2040 [25]. The estimates do not consider only technological readiness but also the capacity gap in aircraft production, the cost of liquid hydrogen and kerosene, infrastructure, and especially the government restrictions on aviation emissions.

Although the first civil aviation applications of hydrogen fuel cell aircraft (e.g. the TU-154) used liquid hydrogen and cryogenic tanks, the first fuel cell aircraft prototypes shifted to pressurized hydrogen in gaseous form. The reasons behind the shift lie in the fact that pressurized hydrogen is cheaper, easier to obtain, easier to handle, and technically easier to implement. Nevertheless, liquid hydrogen has many advantages over gaseous hydrogen in aviation. First, the gravimetric index—a ratio of hydrogen mass and hydrogen-tank system mass—of the pressurized tank is about 5% [26]. In comparison, the gravimetric index of a cryogenic tank can reach up to 70% [27]. Therefore, an aircraft powered by liquid hydrogen has lower weight, better aerodynamic characteristics and consume less fuel, than comparable aircraft on gaseous hydrogen [28]. In addition, the advantage of liquid hydrogen compared to pressurized hydrogen increase as the aircraft fuel requirement for the design mission increases [29]. Therefore, although the first fuel cell prototype aircraft is designed to fly on gaseous hydrogen, future commercial aircraft applications are expected to fly on liquid hydrogen instead.

## 2 Hydrogen Logistics

In general, logistics refers to the process of moving a particular commodity along the supply chain. Logistics involves the managment of various resources (equipment, vehicles, labour, etc.) to ensure that goods are delivered in appropriate quantity, quality, and time at different stages of the supply chain. These commodities can be materials, food, consumables, or, as in our case, hydrogen. In terms of whether the process takes place inside or outside the company we can distinguish between inbound and outbound logistics. While **inbound logistics** mainly focuses on organizing internal processes in such a way that all company processes, be it the production of a certain product or the provision of services, run smoothly, **outbound logistics** deals with the process of providing certain goods from the manufacturer to the end user.

Regarding the process outside the company, outbound logistics provides efficient and effective supply management that links suppliers, distributors, and end users.

Regarding the type of end users, logistics can be further divided into B2B (business to business) when materials, semi-finished products, spare parts are delivered to companies, B2C (business to consumer) when final products are delivered to end users and B2A (business to administration) when logistics services are provided to public authorities (government agencies, municipalities, public companies, etc.). Hydrogen logistics addressed in this context refers to the distribution logistics that manage hydrogen supply at various supply chain stages from one business (hydrogen producer) to the other business (airport) and within airport itself, from on-site production to the aircraft as seen in Fig. 3.

While hydrogen transportation refers only to the hydrogen delivery from the production site to the end-user by finding the most time- and cost-efficient delivery route, hydrogen logistics is much broader and includes complete supply chain management. Besides producing, transporting, storing, and handling, hydrogen logistics also has several other activities that need to be performed when supplying hydrogen. Therefore, hydrogen logistics also refers to, among other things, handling and preparation of hydrogen for transport and further use. This requires several processes to be performed, such as hydrogen compression, liquefaction, and, especially in the case of aviation use, purification to meet the needs of the end-users'. Poorly managed logistics will result in insufficient deliveries, delays, increased costs and consequently loss of business.

## 2.1 Hydrogen Production

Hydrogen cannot be naturally exploited but must be produced by various processes. In general, production methods can be divided into two groups. The first group refers to obtaining and producing hydrogen from raw materials and fossil fuels. Several production methods can be used, such as steam reforming, auto-thermal reforming, partial oxidation, methanol cracking, gasification, etc. The second group

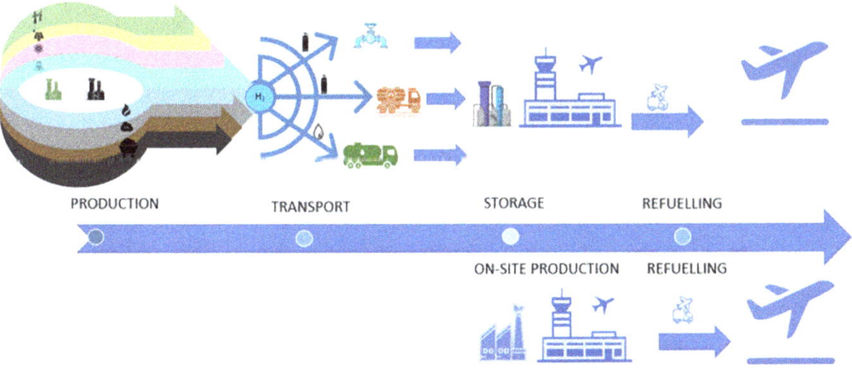


**Fig. 3** Hydrogen logistics to and at the airport (Source: Author, 2020)

refers to obtaining and producing hydrogen from renewable sources and to methods, such as electrolysis (e.g. electricity from renewable sources), biogenic production, dark fermentation, microbial electrolysis cell (MEC), etc. [5]. A standardised colour code is commonly used to differentiate between different methods of hydrogen production and their impact on the environment [30].

Note: The colour codes of hydrogen refer strictly to the production procedure. Hydrogen as a product is always the same colourless element, although it may differ in purity. **Green** hydrogen is produced without greenhouse gas emissions. It is produced by electrolysis from water and electricity coming from renewable energy sources like wind. Nowadays, green hydrogen is produced on a relatively small scale due to its high production price. Only 5% of all hydrogen is produced by electrolysis, partly from renewable energy sources. In the future, the production of green hydrogen will be determined by the price of electricity. If the cost of electricity decreases due to large amounts of electricity from wind or solar power, green hydrogen may become widely available. **Yellow and pink** (sometimes also referred to as purple or red) hydrogen are also produced by electrolysis. For yellow hydrogen, the electricity comes from solar power, for pink hydrogen, electricity is produced in nuclear power plants. Hydrogen produced from methane by steam reforming is **grey** hydrogen. If the carbon capture and storage method is used to capture carbon dioxide formed during the process, the associated hydrogen colour is **blue.** The least environmentally friendly method is the production of hydrogen from black or brown coal (lignite), producing **black** and **brown hydrogen**. Apart from the above-mentioned hydrogen colour code, **turquoise** hydrogen is sometimes used, referring to hydrogen obtained using pyrolysis. Depending on the nature of the process (source of energy, if carbon capture is used or not), it can have different 'shades of green', meaning that it can be more or less environmentally friendly. The term **white hydrogen** is sometimes used to refer to hydrogen that is not produced but occurs naturally. Nevertheless, the exploitation of these natural reserves is not feasible. Also, producing hydrogen biochemically by using specific types of bacteria is still being researched [31]. However, this type of production is still in the experimental phase, so it is unlikely to be able to meet demand in sufficient quantities. There is no colour associated with this kind of hydrogen production. Nowadays, most of hydrogen is produced from methane (68%), but it can also be produced from oil (16%), coal (11%), or other sources like biomass. Centralised hydrogen production is economically more feasible than decentralised [5]. The production of hydrogen by electrolysis is costly and insufficient to supply the economy with the necessary energy [32]. Therefore, for the time being, the most economical method of hydrogen production would be steam reforming, resulting in grey hydrogen. Today, hydrogen can be produced from 10 EUR/kg, while liquefaction adds an additional cost of 1–2 EUR/kg, so the net price of liquid hydrogen should be around 11–12 EUR/kg [33]. Hydrogen is expected to replace natural gas in cases where electrification is not economically not suitable or technically possible (heavy industries and in aviation). Massive investments by the European Commission in renewable hydrogen technology should make it possible to install 40 GW of electrolysers in the EU by 2030, producing up to ten million tonnes of renewable hydrogen annually.

Additionally, the EC is encouraging research and innovation in the hydrogen economy through various programmes and initiatives, such as Horizon Europe, the EU's research programme, Clean Hydrogen Partnership for Europe, an 'Important Project of Common European Interest' for hydrogen, a legal framework for big cross-border projects, etc. [32]. Hydrogen production costs are highhly dependent on natural gas and electricity prices and therefore may change over time. According to some predictions, the price per kilogramme of hydrogen from renewable sources (green hydrogen) may reach 0.85–1.7 EUR/kg and in some cases even be cheaper than blue hydrogen [34].

Regarding the hydrogen production sites, Air Liquide had a large green-hydrogen plant built in Quebec (Canada) in 2021 with the most powerful 20 MW electrolyser to date. Chile plans to build electrolysers with a capacity of 5 and 25 GW by 2025 and 2030 and aims to become the leading exporter of the cheapest green hydrogen. Production sites in Europe, such as in the Netherlands, do not have sufficient resources for renewable electricity, so the current focus is on producing hydrogen with fossil fuels and transforming it to blue hydrogen by storing the $CO_2$ emitted during the production. Scotland has a large volume of natural gas feedstock available as well as massive offshore $CO_2$ storage capacity to provide carbon capture. The country has an ambitious plan to reach 121 TWh production of hydrogen by 2050, which would meet the domestic demand and allow export to Europe. As a first step, Acorn plans to make the first investment to build 200 MW hydrogen production plant at St Fergus gas terminal by 2025 [35]. On the other hand, Spain has enough renewable energy sources, such as wind and solar, and plans to focus on building electrolysers for producing and storing green hydrogen [32].

## Hydrogen Purification

Produced hydrogen can contain impurities such as traces of oxygen, nitrogen, and similar, which means that produced hydrogen is not 100% pure. The degree of purity varies depending on the production method used. For example, hydrogen produced with electrolysis is more refined than hydrogen produced from steam reforming. The number of impurities allowed in hydrogen after production depends on the intended use. Hydrogen intended for combustion engines can be less pure than the hydrogen used in fuel cells. If the purity of hydrogen is too low, it can be further purified using specific procedures. According to ISO/PAS 15594, the purity level of hydrogen for aviation must be at least type II, grade D, with additional requirements are listed in Table 1.

**Table 1** Purity demands for hydrogen use in aviation

| Form | Liquid hydrogen, type II, grade D |
|---|---|
| Purity | >99.9999% (volume fraction) |
| Para-hydrogen | >95% (minimum mole fraction) |
| Total gasses | <100 μmol/mol |
| $O_2$ content | <0.00002% (volume fraction) |
| $N_2$ content | <0.00002% (volume fraction) |
| $H_2O$ content | <0.00005% (volume fraction) |
| $C_nH_m$ | <0.000001% (volume fraction) |
| CO content | <0.000001% (volume fraction) |
| $CO_2$ content | <0.000001% (volume fraction) |
| Total sulfur compounds | <0.004 μmol/mol |
| HCHO content | <0.01 μmol/mol |
| HCOOH content | <0.2 μmol/mol |
| $NH_3$ | <0.1 μmol/mol |
| Total halogenated compounds | <0.05 μmol/mol |
| Particle diameter | <5 μm |

Source: [36]

## Hydrogen Compression

If the hydrogen is intended to be stored in pressure vessels, it needs to be compressed. There are two types of compressors that can be used: centrifugal and reciprocating. The latter is about 50% more expensive but has higher efficiencies and is the most commonly used [37]. Compression costs are depended on the inlet and outlet pressure, and flow rate. Larger compressors have lower unit cost as small compressors. To compress 1 kg of hydrogen, 0.7–1.9 kWh of energy or 2–3% of hydrogen energy content is consumed [38].

## Hydrogen Liquefaction

Hydrogen liquefaction is used when hydrogen is to be stored as liquid hydrogen in a cryogenic tank. The most common methods for liquefaction of hydrogen are Linde's and Claude's cycle. During liquefaction, particular attention needs to be paid to ortho- to para-hydrogen conversion, usually by using a catalyser to accelerate ortho- to para-hydrogen conversion. Namely, the hydrogen atoms in a molecule can spin either in the same direction (ortho-hydrogen) or the opposite direction (para-hydrogen). At average temperature and pressure, ¾ of all molecules are in the form of ortho-hydrogen and ¼ in the form of para-hydrogen. In a liquid form, where temperatures drop below 20 K, most hydrogen (99.8%) is in a para-hydrogen form. The conversation of ortho-hydrogen to para-hydrogen can take several days and releases a considerable amount of energy. Therefore, if hydrogen is liquefied too quickly and stored before conversion to para-hydrogen occurs this will result in boil-off [39]. Liquefaction of hydrogen is a complicated and energy-consuming

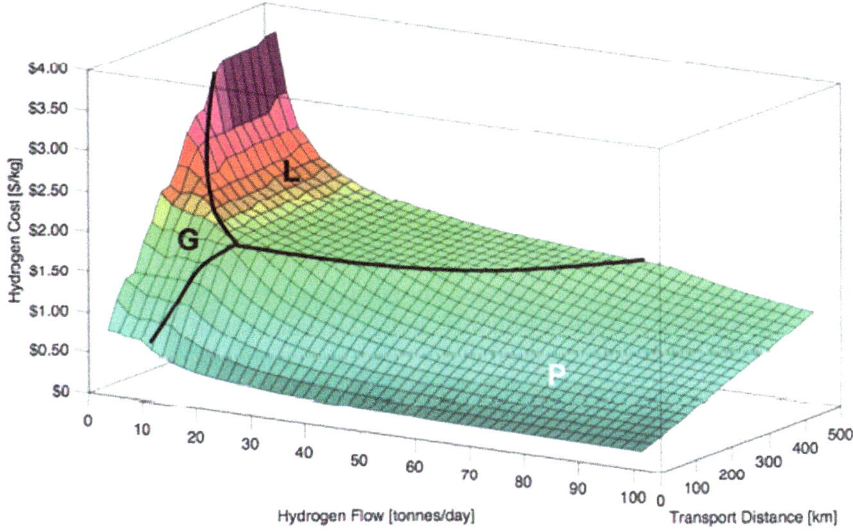


**Fig. 4** Hydrogen delivery modes depending on volumes, hydrogen flow rates, and distance [33]

process: 11 kWh/kg of energy is required to liquefy 1 kg of hydrogen, equal to 33% of base energy content [33]. It is not feasible to liquefy small quantities of hydrogen. The liquefaction plants should be designed so that a flow of at least 20 tonnes of hydrogen per day can be expected as shown in Fig. 4 [33].

## 2.2 *Hydrogen Transport*

Transportation refers to either the delivery of goods or the movement of people by different transport modes, such as road, rail, water, sea, cable, pipeline, air, including the infrastructure, vehicles, and operations. Hydrogen can be delivered to the airport by road, rail, water, or pipeline.

### Transport by Road

Depending on its form, hydrogen can be transported by road, either in a tube trailer or a $LH_2$ trailer. When hydrogen is transported by a tube trailer, hydrogen is compressed at the production site and delivered to the customer site by the tube trailer trucks. The trailer is dropped off at the delivery site and picked up when it is emptied. The majority of costs associated with hydrogen transport with tube trailer are related to the operating cost of the truck such as expenses for maintaining truck and labour costs. As the distance increase the cost of such delivery increases. Therefore, such delivery is the best option for delivering small quantities of over short distances [33]. Nowdays, this delivery option is particularly interesting for refueling small prototype fuel cell aircraft, since it does not require additional airport infrastructure. Nevertheless, since future aircraft are expected to use liquid hydrogen as fuel and it is unfeasible to liquefy small quantities of hydrogen, this delivery option will be replaced by other modes of transportation in the future.

If hydrogen is transported by road using $LH_2$ trailer, the hydrogen must first be converted from gaseous to liquid form (e.g. liquefied) at the production site and transported to the customer by a $LH_2$ trailer. $LH_2$ trailers have a range of about 4000 km. Over longer distances, hydrogen heats up, increasing the pressure in the container to the point that the hydrogen must be vented. The largest cost component when transporting hydrogen with $LH_2$ trailer, is liquefaction, which can reach between 80% and 95% of total cost of delivery. When delivering liquid hydrogen, a travel distance will not significantly impact overall delivery cost [33]. Due to the higher density of liquid hydrogen compared to gaseous hydrogen, a larger amount of hydrogen can be transported with an $LH_2$ trailer than with a tube trailer. At a density of 70.8 kg/m$^3$, around 3500 kg of liquid hydrogen can be carried at a loading volume of 50 m$^3$. Like trailer transport, $LH_2$ can also be transported by rail, given that suitable railway lines, and loading terminals are available [40]. Delivery by $LH_2$ trailer is feasible for relatively small quantities of hydrogen over large distances (Fig. 5) [33].

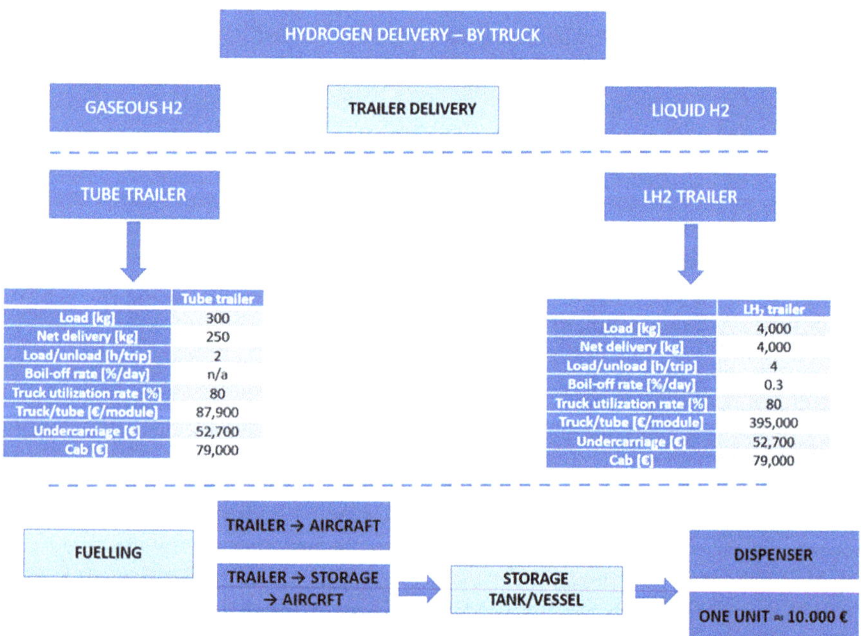


| | Tube trailer |
|---|---|
| Load [kg] | 300 |
| Net delivery [kg] | 250 |
| Load/unload [h/trip] | 2 |
| Boil-off rate [%/day] | n/a |
| Truck utilization rate [%] | 80 |
| Truck/tube [€/module] | 87,900 |
| Undercarriage [€] | 52,700 |
| Cab [€] | 79,000 |

| | $LH_2$ trailer |
|---|---|
| Load [kg] | 4,000 |
| Net delivery [kg] | 4,000 |
| Load/unload [h/trip] | 4 |
| Boil-off rate [%/day] | 0.3 |
| Truck utilization rate [%] | 80 |
| Truck/tube [€/module] | 395,000 |
| Undercarriage [€] | 52,700 |
| Cab [€] | 79,000 |

**Fig. 5** Hydrogen transportation by road. (Source: Author based on [33])

## Transport by Pipeline

Hydrogen can be transported through pipelines similar to natural gas. Hydrogen must first be compressed to around 24–130 bars at production site, before it can be delivered through the pipeline. Because hydrogen can easily leak, penetrate through, and embrittle materials, a unique pipeline system needs to be built to prevent this. Current systems consist of steel and carbon steel pipes such as mild strength steel—API 5L X42 or X52 and ISO 13847 which are corrosion resistant and have automatic relief valves. While the initial construction cost of hydrogen pipeline system is very high, once built it represents the cheapest way of hydrogen transportation.

Pipeline systems can stretch for hundred and hundred kilometres. In Europe, the longest pipeline extends to 613 km in Belgium, while other countries such as Germany, France, and the Netherlands also have pipeline system that average about 300 km in length. They are built at a minimum depth of 90–120 cm with pipes 10–30 cm in diameter. The amount of gas that can be delivered through pipeline system is very much dependent on the length of the system, volume flow, and pressure. The pipeline can also be used as storage system by adjusting the pressure to better handle on one side supply and on the other side demand [41]. The main cost of this type of delivery of hydrogen will be on the side of initial investments necessary to place pipeline installation into space and environment. This cost varies from country to country and from location to location, either is a developed area with all supporting infrastructure or rural area. As seen from Fig. 6, the initial capital investment in various projects from 1993 to 1997 for pipelines ranged from €270,000 to €810,000 per kilometre [42].

For pipeline delivery, the cost to compress, storage and dispense hydrogen, will be somewhere between 1.72 and 2.41 €/kg, with the most likely cost of 2.06 €/kg of hydrogen. Since the most critical cost component in transport by pipeline is capital cost of pipeline, this transportation method is most suitable for transporting large quantities of hydrogen [33]. In the 1970s, NASA explored different hydrogen delivery options to airports operating commercial flights, depending on the hydrogen

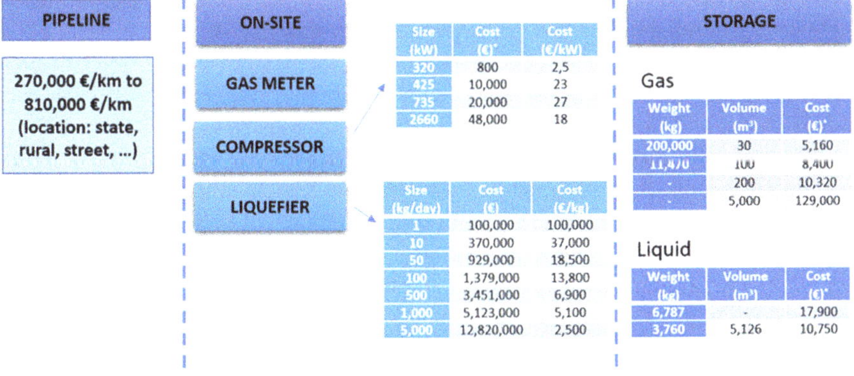


**Fig. 6** The cost of a hydrogen pipeline. (Source: Author based on [41])

form and delivery distance. The study concluded that for short distances (i.e. less than 64 km) and large demand, transporting gaseous hydrogen with pipelines and liquefy it in the vicinity of the airport was the most economical option, while for longer distances, transportating liquid hydrogen by rail or truck would prevail [43].

### Transportation by Water

Hydrogen can be transported by water in suitable containers (e.g. ships). In this case, the hydrogen should be kept at −253 °C to keep it in liquid form and avoid the risk of damage to the containers. Although such transport is theoretically possible and several conceptual solutions have been outlined, it has not yet become a common practice due to a lack of massive demand for such transport, which drives up the delivery costs and makes it economically unfeasible.

At the moment, there are only a few pilot projects developing ships for the transport of hydrogen in Europe and Asia in the coming years. In 2017, Norwegian shipping company Wilhelmsen began developing the first Liquid Hydrogen (LH2) bunker vessel BV Tomorrow. Wilhelmsen has been working with Moss Maritime, Equinor and DNV-GL to develop LH2 bunker vessel. BV Tomorrow is a concept design for an LH2 Bunker Vessel capable of loading, transporting, and discharging liquefied hydrogen to seller vessels and receiving terminals, with a cargo capacity of 9000 $m^3$ of LH2 or 500 tonnes [44].

The world's first ship for transporting hydrogen by water (e.g. Suiso Frontier) was built in Japan by Kawasaki Heavy Industries. The prototype ship is undergoing sea trials and is scheduled to make a 9000 km demonstration voyage from Australia to Japan in 2021. Suiso Frontier is relatively modest, measuring 116 meters in length and weighing 8000 gross tonnes. On its maiden voyage, it will run on diesel, but the company plans to use hydrogen to power future, larger commercial vessels. The 1250-cubic-metre tank for the hydrogen is double-shell and vacuum-insulated to maintain temperature [45]. In this pilot programme, the hydrogen will be produced in Latrobe Valley (Australia) using a lignite gasification process that would produce about 3 tonnes of hydrogen from 160 tonnes of coal. This would then be delivered by trailers from the production site to the port of Hasting (e.g. about 150 km away), from where the Suiso Frontier will travel roughly 4893 nautical miles to Kobe (Japan) [46]. The plan is to build a commercial-scale hydrogen transporter by the mid-2020s and have it in service by 2030 [45].

Another pilot project is underway in South Korea, one of the world's largest shipbuilding hubs. Korea Shipbuilding & Offshore Engineering (KSOE) is the first company in the country to work on building a commercial liquid hydrogen carrier. The company plans to submit the first international standard for hydrogen ships to the International Maritime Organization (IMO) by 2022. In order for ships to operate worldwide, their design should be based on the IMO's shipping regulations. To date, however, there are no standards for hydrogen ships. KSOE and Korean Register (KR) will evaluate the settings for safe handling of hydrogen based on advanced technology, which includes cargo handling systems, gas storage on ships and for fuel supply systems [47].

## 2.3 Hydrogen Storage

Typically, hydrogen is stored as a compressed supercritical fluid at room temperature (in pressure vessels) or in a liquid state under normal pressure (in a cryogenic tank), while also other storage options could apply such as cryogenic tanks for slush (solid-state) hydrogen or as cryo-compressed hydrogen.

In pressure vessels, hydrogen is usually stored under a pressure of 350 or 700 bar. The volumetric energy density of hydrogen compressed to 350 bar is 2.9 MJ/L, while the volumetric energy density of hydrogen compressed to 700 bar is 4.8 MJ/L. Pressure vessels holding hydrogen at 700 bar are usually made of carbon fibre or other composite materials [26]. The capital costs for pressure vessels ranges from 26 to 172 €/kg hydrogen [42].

Cryogenic tanks store hydrogen at an average pressure and temperature of about −250 °C. Cryogenic tanks consist of several layers with vacuum or unique insulation materials in between. Since hydrogen slowly evaporates, the pressure inside the cryogenic tank increases over time. This reaction is called boil-off. Therefore, cryogenic tanks are equipped with a vent that releases hydrogen into the atmosphere when the pressure inside the tank reaches the permitted limit. How long hydrogen can be stored in cryogenic tank before venting is required depends mainly on the tank specifications. Depending on the intended use, tanks can be designed for both short and long periods before venting is needed. Due to a better surface-to-volume ratio, large hydrogen storage units can handle boil off more easily and are therefore less expensive. In several projects carried out from 1993 to 1997, the costs for storage ranges from 27 to 610 €/kg [42].

Apart from these two most common storage methods, hydrogen can be stored in gaseous form under average temperature and normal pressure inside underground caverns. For example, these caverns can be old, abandoned mines or natural caverns. Hydrogen can also be stored in cryogenic tanks as slush (solid-state) hydrogen or as cryo-compressed hydrogen, where hydrogen is compressed and cooled simultaneously. Slush and cryo-compressed hydrogen can reach relatively high volumetric energy densities, but unfortunately, storage costs are high and usually economically unfeasible. Hydrogen can also be stored as inside materials and material-based $H_2$ storage systems. Hydride storage is the most common method, where hydrogen is absorbed inside a metallic lattice of metals such as palladium, magnesium, or aluminium [5, 28].

## 3 Application of Hydrogen at Airport

The transition to hydrogen in aviation will depend on many aspects, such as the future hydrogen production costs, technology readiness level of hydrogen-related technologies for production and aircraft, possible environmental restrictions on aviation, price of jet fuel, necessary airport infrastructure, etc. In order to use hydrogen aircraft for commercial flights, sufficient number of airports must build facilities for their operation and fuelling [25]. Currently, no airport holds appropriate facilities for fuelling hydrogen aircraft, as they are not yet used in commercial flight operations. However, there are already some facilities at airports that allow refuelling of ground vehicles with hydrogen.

### *3.1 Aircraft Handling and Refuelling Procedure*

To date, there, is no established procedure for refuelling hydrogen aircraft. Only ISO 15594:2004 [36] explicitly addresed there fuelling of hydrogen aircraft at airports, and was withdrawn in August 2019. Still, it highlights some important aspects to consider when using hydrogen as an aviation fuel.

According to ISO 15594:2004 [36], an airport should provide two connection points between the aircraft and ground support: a refuelling connection to supply liquid hydrogen to aircraft tanks and a boil-off connection points to release gaseous hydrogen from aircraft tank due to boil-off. Due to safety reasons, these connection points should be in open space, free of flammable and combustible objects (e.g. trees), well away from other facilities and should have restricted access. Aircraft must be properly grounded and bounded before beginning operation. At the refuelling point, aircrafts should provide three types of service: normal refuelling during aircrafts' turnaround between flights (a cold system fuelling), de-fuelling due to planned maintenance and troubleshooting, and first refuelling of new aircraft or refuelling an aircraft after maintenance and troubleshooting (warm system refuelling). This includes procedures such as purging with an intermediate gas (e.g. helium) and precooling and warm-up procedures for the aircraft tank and connecting hose and coupling.

During regular refuelling, personnel should first check that the tanks are cold and still contain at least a small quantity of liquid hydrogen. Next, they should preform purging and precooling of connecting hose. During refuelling, boil-off valves of a tank should be opened, and the fuel level of the tank is monitored. After the tank is filled with fuel, boil-off valves are closed, and aircraft is disconnected from the connection point. During warm system refuelling, in addition to normal refuelling, the tank should first be purged with intermediate gas to remove air or other gasses from the tank. The tank should then be precooled with conditioned hydrogen which should also push out the inert gas from the system.

Stationary storage of fuel at the airport is neither requried nor recommended—fuelling should be done directly from portable storage, in which hydrogen was transported to the airport (e.g. truck). At the aircraft interface refuelling point, the temperature of the liquid hydrogen should be 20 K or lower. The pressure should be higher than 700 kPa to achieve good fuelling times (20 min) [36]. The refuelling point on aircraft may be located in the tail or nose of aircraft to provide more safety in case of unsecured spark ignition [43]. A refuelling coupling unit for a small aircraft can be manual and must include a refuelling hose, a refuelling connector, and safety monitoring equipment. A refuelling connector, together with the attached part of refuelling hose, shall not exceed 10 kg (or preferably 7 kg). The connector should have a diameter of 30 mm to meet the requirements of the connectors used for road vehicles. The system should have automotive sensing and shut down option [48] and include a philtre to tra particles larger than 5 micrometres. A philtre should be detachable and cleanable. Safety equipment should include battery powered monitoring devices to measure pressure, temperature, flow rate, filling level of the tank, hydrogen leak, and valve position and a portable detector of hydrogen concentration and heat [36]. All monitoring systems should be connected to an interface, be easy to operate and respond automatically in the event of failure. Personnel should be alert for possible deterioration of thermal insulation. Later, this can be detected by observing possible water condensation or water freezing on a tank or system surface.

A safety area should be designated and the accumulation of hydrogen vapours in enclosed spaces should be prevented to avoid a potential fire hazard. In this context, a precautions should also be taken to control large quantities of LH2 in critical areas such as fuel facilities [48]. In case of hydrogen plant operation and installation of infrastructure, fire protection aspects have already been widely addressed by the National fire protection association [49]. With minimal requirements, the boil-off hydrogen can be released directly and safely (open environment). Nevertheless, equipment for re-catching hydrogen should be preferred for economic and safety reasons [36].

### *3.2 Application of Hydrogen for Airport Handling Ground Vehicles*

Most of the existing hydrogen applications at airports are for refuelling ground vehicles, especially ground support equipment (GSEs) that causes a significant portion of emissions during their operations [50]. The International Civil Aviation Organization (ICAO) suggests measures that should be considered to reduce the fuel consumption of GSEs and associated indirect emissions, such as replacement of GSE engines, new diesel engines, use of electric vehicles or alternative fuels such as compressed or liquefied gas; the expected reduction in fuel consumption is 8–18% [51]. In the last decade, hydrogen and fuel cell technologies have been explored as one of the possible solutions to mitigate pollution. Hydrogen has already been introduced as refuelling option for ground vehicles at several airports such as Munich [52], Vancouver [53], Berlin Brandenburg [54], Paris Charles de Gaulle [55], Liège [56], Gatwick in 2019 [48], and most recently Bremerhaven [57] and Seoul Incheon International airport [58].

At Munich airport, a two parallel hydrogen supply paths were introduced, one for refuelling three buses with gaseous hydrogen (GH2) and one for refuelling passenger cars with liquid hydrogen (LH2) during the period 1997–2000. While GH2 was produced locally mainly by steam-methane reforming and partly by high-pressure electrolysis, the liquid hydrogen was delivered by trailers from a nearby liquefaction plant and stored in the station's LH2 storage tank. The LH2 filling station was located outside the closed airport areas accessible to the public [52].

At Vancouver airport hydrogen was used for buses and tractors for baggage carts equipped with fuel cells in 2006. At St. Louis airport, hydrogen was used for service vehicles such as forklifts, on police personnel carriers and in a minibus for passenger or crew transport, as well as for a backup and emergency system. Given the limited amount of hydrogen needed, the airport produced all the gaseous hydrogen needed using small steam methane reformers [53].

At Berlin-Brandenburg Airport, the first ground-based fuel cell electric vehicles (e.g. buses and cars) were refuelled with gaseous green hydrogen produced on-site by electrolysis from wind and solar energy as part of the 2014 demonstration project. The hydrogen plant was the first alkaline high-pressure electrolyser on the market capable of, producing 200 kg of hydrogen per day to fuel 50 vehicles. Two hydrogen fuelling station were set up in coexistence with natural gas, LPG, and electricity refuelling stations [54].

At Paris-Charles de Gaulle airport, Air Liquide set up a new hydrogen station to power ground vehicles, while a larger network of hydrogen refuelling stations was developed for Paris metropolitan area. The Paris taxi fleet 'Hype', the world's first ever hydrogen-powered taxi fleet, was developed in 2015 by the start-up (Société STEP du Taxi Électrique Parisien) in collaboration with Air Liquide. The fleet consist of more than 50 hydrogen-powered vehicles and is expected to grow to 600 taxis by 2020 [55].

Liège airport was equipped in 2018 with facilities for the production, distribution, and use of green hydrogen. The hydrogen is produced through use of electricity gained by solar panels. Hydrogen distribution stations on the airport site were used to supply the airport's own fleet as well as to supply other vehicles and to create a hydrogen cluster [56].

The most recent Bremerhaven Airport, built in 2020 an electrolysis hydrogen pilot plant with the wind turbine, that can supply the energy required for the hydrogen fuel cell with a capacity of up to 1000 tonnes per year. This will be enough to power 200 cars with a range of 600 km [57]. In the same year, hydrogen mobility solutions, including fuel cell buses, hydrogen stations, and hydrogen supply, were offered at Seoul Incheon International Airport in a colaboration between Air Liquide Korea, Incheon International Airport Corporation, Hyundai Motor Company, and

HyNet (Hydrogen Energy Network). Two high-capacity hydrogen filling stations, the largest to date, have been installed [58].

## 3.3 Application of Hydrogen at Small- and Medium-Sized Airports

Based on a study by Clean Sky 2 JU et al. (2020) smaller airports would be a good starting point for gradual adoption of H2 for aviation due to their low air traffic and less congestion and space limitations. Considering the Trans European Airport Network, we can classify airports into major airports or hubs and secondary airports, whether they have more or less than five million travellers per year, while airfields refer to airports for leisure, training, and sports. On the total 3029 airports, 69 are considered hubs, 1908 are secondary airports, and 1101 are airfields. For secondary airports, the average runway length is 996 m, with minimum and maximum lengths of 108 m and 4000 m, respectively. Most secondary airports have runways of 600 m or longer. Of all secondary airports, 44% have either concrete or asphalt runways, while 48% have a grassy surface. Less than 3% of all airports have runways of other types of surfaces like gravel, soil, and sand [59]. When referring to small- and medium-sized airports, we have in mind land airports with little or no scheduled service and light general aviation traffic. Such airports have the greatest potential for adoption of hydrogen-powered airplanes. Considering small-sized, medium-sized, and closed airports in the EU as potential airports for implementing hydrogen-powered aircraft, they would require at least an 800 m runway, with concrete, bituminous, or asphalt surface, located in the vicinity of an existing hydrogen manufacturer or H2 fuelling station. There are 7277 airports in the EU that can be categorised as closed, small-, medium-, or large-sized airports. There are 525 medium, small, and closed airports (no heliports) with IATA codes that have a runway of at least 800 m in length, and are paved with asphalt, bituminous surface, or concrete (see Fig. 7) [60].

Hydrogen logistics and especially transportation are heavily influenced by the distances between the facilities where hydrogen is produced and delivered to end-users, i.e. airports. Therefore, distance plays an important role, influenced by the time and cost of hydrogen delivery. In Europe, there are 163 different locations of hydrogen producers. By comparing the locations of hydrogen producers and airports, the following can be concluded:

- At distance of 100 km from the hydrogen producers there are 128 airports (10 closed, 12 small, and 106 medium)
- 150 km: 77 airports (5 closed, 8 small, and 64 medium)
- 200 km: 65 airports (4 closed, 5 small, and 56 medium)

For smaller airports where smaller quantities of hydrogen are required, delivery by $LH_2$ trailer would be a more economical option than on-site production, which

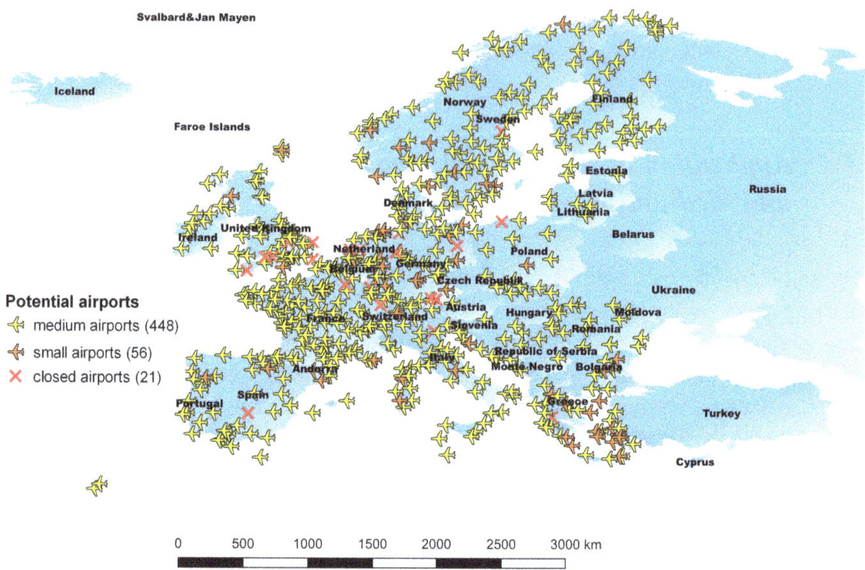


**Fig. 7** EU small-sized, medium-sized, and closed airports with IATA code and 800 m + runway. (Source: Author based on data retrieved from (Our airports, 2020))

involves very high on-site liquefaction costs. For small and medium sized airports that provide feeder and regional flights, it is assumed that no significant changes to runways, taxiways, terminals, and stands would be required. Refuelling directly from the trucks will require reorganizing traffic at airports and organizing refuelling while considering the relevant procedures and safety aspects.

The main disadvantage in supplying small airports is the distance between airports and hydrogen producers. In Europe, there are 255 airports with 1–70 seats aircraft traffic that they are located more than 200 km away from the hydrogen producer (see Fig. 8). There are 31 small airports, 22 medium airports, and 2 closed airports. As we can see, most of the airports are in Scandinavia, western part of France and islands, Greece, Romania, Portugal and other countries. These airports might have difficulties to get hydrogen supply because of their distance from hydrogen producers, as it might become impracticable or just to costly.

A 19-seat fuel cell aircraft would require about 200 kg of hydrogen for flights with a range of 500 km and a 70-seat fuel-cell aircraft would require about 700 kg of hydrogen for the same distance. If considering existing daily flights in this aircraft segment and introducing a hydrogen aircraft, almost all airports operating 19-seat aircraft (e.g. 96%) and half of the airports operating 70-seat aircraft would need to provide less than 1 tonne of liquid hydrogen per day, which would be sufficient for daily hydrogen aircraft operations. For smaller airports, that have higher volume of flights this would also result in a higher demand for hydrogen. This is the case of Tromsø airport, where 44 tones of hydrogen would need to be provided daily for operation of 19-seater and 70-seater hydrogen aircraft [28].

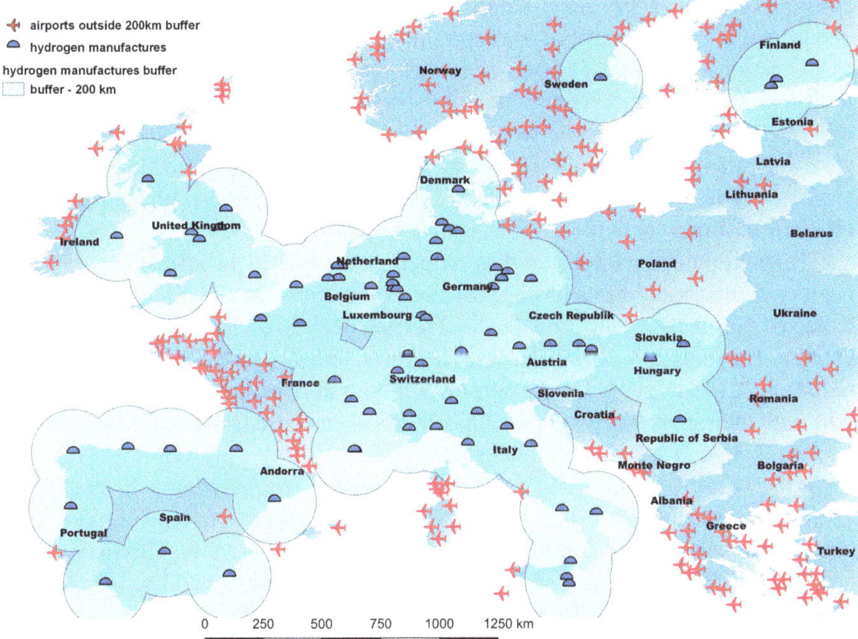


**Fig. 8** Airports outside 200 km buffer zone of hydrogen manufacturer. (Source: Author based on data retrieved from (Our airports, 2020))

## *3.4 Application of Hydrogen at Large Airports*

Since there are no practical applications at airports yet, some theoretical studies have been conducted on the requirements for hydrogen aircraft operation at major airports, such as for the case of Los Angeles airport [61] and for the case of London Heathrow airport (UK) [9]. Given that the transition to hydrogen powered aviation is going to be gradual, this will require from airports to provide facilities and ground vehicles for fuelling and serving both hydrogen and kerosene aircraft [25]. The pace at which the fuelling facilities for hydrogen aircraft will be introduced at airports will depend on the extent to which airliners incorporate hydrogen aircraft into their fleets. Therefore, in the near future, it is more likely that the refiling from hydrogen plant will be provided by trucks, mobile refuelling stations, or refuelling slots located away from space for embarking passengers [6].

As hydrogen aircraft would gradually replace conventional aircraft and hydrogen demand would increase, hydrogen production and liquefication can be done on-site or near-site. Since on-site production will not be possible anytime soon, it is expected that hydrogen production facilities will be based on producing green hydrogen. One of the studies on this matter was conducted for Los Angeles airport, where hydrogen would be produced through electrolysers and use of wind turbines [61]. The most recent study on integration of hydrogen aircraft in the air transport system also considers overall supply chain, safety, and cost considerations [48]. The latter study underlined the opportunities for airports in using hydrogen as an aviation fuel and as an energy source for all airport activities (e.g. heating, lighting, cooking, ground transportation vehicles). As aircraft tanks should be as light as possible, they will be designed to sustain hydrogen without boil-off only for short periods of time sufficient for turnaround. During periods of inactivitye (e.g. during the night), the aircraft would vent hydrogen, which could be captured and used as energy source for all kinds of airport activities.

The environmental impact of the production site near to the airport should not be neglected and should be as low as possible. Although noise and air pollution can be expected during the production and liquefaction of hydrogen, this is negligible compared to the impact of air traffic. With regard to exhaust gases, the main issue is to reduce $NO_x$ levels. As far as the approval of normal operation is concerned, an airport with hydrogen production facilities might even cause fewer environmental and legal problems than a conventional airport [62].

In the case of on-site production, airport spatial requirements will need to be examined, ensuring that production is safely located away from other airport premises and runways and protected by hills or other physical barriers. This may require additional infrastructure investment in airports, in addition to building a production plant and providing delivery infrastructure, trucks, or even a pipeline system for hydrogen distribution. To cope with the dangerous situations, firefighting equipment and locations at airport may need to be changed to ensure that firefighters can reach the scene quickly enough due to the short duration of the LH fire [48]. Moreover, the current airport infrastructure is adapted to the physical characteristics of the current aircraft, raising the question of whether the newly designed hydrogen aircraft will be compatible with the existing infrastructure. If the size of hydrogen aircraft increases, this may affect the length and width of runways and taxiways. At the same time, the increased size (longer wingspan and fuel storage) means that the terminal building and gates will need to be redesigned [48]. As some conceptual designs of hydrogen-powered medium- and long-range aircraft assume they will have 10–15 m of additional length, this would lead to necessary infrastructure investments for resizing the airport gate [6] and lengthening the passenger boarding bridges [9]. The turnaround time could also increase compared to conventional aircraft if the facilities for parallel refuelling and other activities such as catering and boarding are not available [48] or if refuelling lots are not installed close to the gates or gate space [6]. The increased turnaround rate leads to the under-utilization of existing infrastructure and necessitates expansion [48]. If aircraft refuelling of aircraft were to be carried out by trailers, that may cause additional congestion at airports, necessitating reorganization of airport traffic and new types of docking and refuelling facilities that should be available at every airport stand. In some cases, airport infrastructure will need to be connected to the hydrogen grid system.

Since production and storage are highly dependent on the amount of hydrogen needed and especially on peak demand, it would be best to spread the production schedule over several modules to expand or reduce capacity and storage as needed. The storage capacity must be sufficient not only to store the production volumes but also to have a reserve in case of production failure. If hydrogen is to be used at large airports, and about 500 tonnes of liquid hydrogen were required, the production facility that could supply such quantities would need 25,000 $m^2$ of space to accommodate the liquefier and storage, and would cause a very small fraction of the footprint (0.2% of the footprint that Heathrow Airport creates today) [6]. The hydrogen storage should be stored away from airport facilities, traffic, and kerosene [62]. As airports offer more and more flights, the hydrogen supply will also need to increase, so the infrastructure will need to be scaled up [48]. In this regard, sharing hydrogen supply infrastructure with other modes of transport could reduce the costs for aviation [25]. This may also refer to recent ideas of developing airports as hydrogen hubs [63]. There are some interesting approaches to transform airports into hydrogen hubs for the region, and such is the case of France, where, through a French recovery plan, substantial amounts are being devoted to develop a disruptive solution throughout the entire hydrogen supply chain between 2020 and 2030 [63].

## 4 Safety Measures

While the hydrogen industry has established procedures for handling hydrogen, these procedures have yet to be adapted for aviation to develop the necessary strategies for operations with hydrogen at airports and specially to address specific hazards. Today, several standards already exists for hydrogen production, use, transportation, and refuelling, primarily for road transportation applications. Existing standards, practices, and regulations can give valuable insights into the handling of hydrogen in a somewhat different environment, such as the airport itself. There are about 150 ISO standards addressing various aspects regarding hydrogen, 83 of which relate to use of hydrogen as a fuel, for road transport, refuelling, storage, generators, dispensing, purification, etc. About 36 new standards are under development, mainly related to the use of compressed gaseous hydrogen for road vehicles as well as for fuel cell applications. In contrast, the use of liquid hydrogen applies only to ships and other marine technology [64]. Some of the existing standards that could also be relevant for aviation are listed in Table 2.

One of the major challenges in handling hydrogen is safety measures. Hydrogen is a highly flammable gas and because of Hindenburg disaster in 1937 [65], there are still fears and doubts about hydrogen safety among general public. Nevertheless, in recent years, we learned a lot about the characteristics of hydrogen and can better control its usage. Basic safety considerations of hydrogen systems are described in ISO 15916:2015 [39]. Being the lightest element in universe, it possesses higher buoyancy and diffusivity than other gasses. This means that hydrogen tends to disperse and form an ignitable mixture with air. In an unenclosed space, the mixtures dilutes below a lower flammability limit, but this cannot happen very quickly and depends on weather conditions. Hydrogen evaporating from the cryogenic tank is denser and therefore will dilute slower, which means that there is a risk of leakage over a long period of time. Hydrogen also has a low viscosity, which leads to high

**Table 2** Relevant standardization

| Standard id. nr. | Standard full name |
|---|---|
| ISO 13984:1999 | Liquid hydrogen—Land vehicle fuelling system interface |
| ISO 15594:2004 | Airport hydrogen fuelling facility operations |
| **ISO 13985:2006** | Liquid hydrogen—Land vehicle fuel tanks |
| ISO 16110-1:2007 | Hydrogen generators using fuel processing technologies—Part 1: Safety |
| ISO 14687-2:2012 | Hydrogen fuel—Product specification—Part 2: Proton exchange membrane (PEM) fuel cell applications for road vehicles |
| ISO 13847:2013 | Petroleum and natural gas industries—Pipeline transportation systems—Welding of pipelines |
| ISO 15916:2015 | Basic considerations for the safety of hydrogen systems |
| ISO/TS 19883:2017 | Safety of pressure swing adsorption systems for hydrogen separation and purification |
| ISO 16111:2018 | Transportable gas storage devices—Hydrogen absorbed in reversible metal hydride |
| ISO 21029-1:2018 | Cryogenic vessels—Transportable vacuum insulated vessels of not more than 1000 litres volume—Part 1: Design, fabrication, inspection, and tests |
| ISO 14687:2019 | Hydrogen fuel quality—Product specification |
| ISO 19880-5:2019 | Gaseous hydrogen—Fuelling stations—Part 5: Dispenser hoses and hose assemblies |
| ISO 20421-1:2019 | Cryogenic vessels—Large transportable vacuum-insulated vessels—Part 1: Design, fabrication, inspection, and testing |
| ISO 17268:2020 | Gaseous hydrogen land vehicle refuelling connection devices |
| ISO 19880-1:2020 | Gaseous hydrogen—Fuelling stations—Part 1: General requirements |

Source: ISO (2021) [64]

flow rates in the case of leak. Due to its low buoyancy, when released, hydrogen will rise to the upper atmosphere, where it would either oxide to water or react with pollutants or escape into space. If comparing leak of liquid hydrogen with kerosene or other conventional liquid fuels, it can cause more safety issues but causes less environmental problems as it cannot seep into the ground, and so it does not pollute surface or ground water or soil [62]. Therefore, escaped hydrogen does not pose a threat to environment.

Hydrogen also responds rapidly to heating and cooling, resulting in rapid changes in temperature and pressure. Moreover, liquid hydrogen expands 23 times more than water when heated under ambient conditions. Therefore, a special care must be taken when filling cryogenic vessels to leave enough room for expansion. Also, due to conversion from ortho- to para-state or due to increased heat exchange with the environment (e.g. due to insulation damage), significant evaporation of liquid hydrogen can occur inside the tank causing sudden rise of pressure. Therefore, the tanks must be equipped with the safety valves that can release hydrogen gas due to boil-off in the case that the pressure in the tank rises over the allowed limit.

The most famous property of hydrogen is its high flammability. Hydrogen can ignite in both low and high concentrations with oxygen from approximately 4% to 77% volume. This is the largest ignition range, if compared with other flammable substances (methane, propane, diesel, petrol). The lower ignition limit is similar to the lower ignition limit of other flammable gasses, but its upper ignition limit is much higher. This means that other flammable substances will not ignite at rich mixtures (low concentration of oxygen and high concentration of flammable substances), but hydrogen will. Therefore, vessels containing hydrogen must be purified with inter gas to remove the presence of air. Hydrogen flame is usually invisible, so special equipment must be used to detect it. Hydrogen can also explode, deflagrate, and detonate, causing serious damage. Because hydrogen can easily ignite not only by open flame but also by hot surfaces, friction, or static sparks below the threshold of human sensation, hydrogen must always be separated from the oxidizer (oxygen and air).

Another important property of hydrogen is the way it reacts with certain metals. Being a light and small molecule, it can penetrate the metal crystal structure of metal. This can cause a significant loss of ductility in certain metals when they are exposed to hydrogen. This phenomenon is known as embrittlement. Also, at temperatures above 200 °C, a non-reversible degradation of steel microstructure can occur, a phenomenon known as hydrogen attack.

Summarizing, hazards that can be caused by hydrogen can be divided into three major categories: ignition hazards including all damage caused by fire, explosion, deflagration or detonation, hazard related to degradation of materials in contact with hydrogen like hydrogen embrittlement and hydrogen attack and health hazards like cold and hot burns and asphyxiation. The low temperature of liquid hydrogen do not usually hurt the skin, while injures may occur in case of touch of metal parts. Hydrogen as such is not a toxic gas, but like carbon dioxide, it can displace oxygen and cause suffocation due to lack of oxygen.

Considering hydrogen characteristics and hazards it can cause, several safety measures can be suggested. First, the amount of hydrogen involved in operation should be minimised. The pipeline diameters and operational pressure must be as low as possible and just enough to satisfy requirements for mass flow. If hydrogen is stored at airports, storage capacities should be as minimal as possible for normal operation. Moreover, whenever possible, operations with hydrogen should be done in an open space. This way, if hydrogen leaks, it will rise and disperse, eliminating the danger of asphyxiation hazard and reducing the possibility of ignition. If hydrogen is present in buildings, non-combustion materials must be used, ignition sources must be avoided, and the building has to be adequately vented all the time. It should be considered that if any hydrogen leakage is present, hydrogen will accumulate at the ceiling. Ignition sources (e.g. ceiling lights) at the top of the building must be avoided. Special equipment designed, fabricated, and tested according to standards and regulations for hydrogen must be used. Equipment must be constructed from appropriate materials that do not react with hydrogen (embrittlement, hydrogen attack, resistance to hot/cold temperatures). Materials must be subject to low stress and have an approved venting system, pressure control system (safety valves), and proper labelling. Hydrogen must be isolated from oxidizers (air) and ignition sources. Vessels and pipes designed to contain hydrogen must be purified with inert gas (helium) and ignition sources must be identified and avoided. Vessels must be kept under positive pressure to prevent air intrusion. If hydrogen must be disposed of (safety valves), the release must be in a safe environment and, if possible, at a height above other facilities.

All equipment shall be properly maintained and inspected according to well-defined procedures and protocols. Equipment shall be inspected regulary, cleaned appropriately and periodically warmed up and purged. All filters shall be cleaned and replaced on regular basis. All evacuation routes must be adequately maintained and free of obstacles. Well-defined safety protocols and procedures must be set. Hydrogen facilities should be separated from other facilities and people. The area around hydrogen facilities where ignition, deflagration, or detonation of hydrogen can cause damage should be identified. This area should have restricted access and properly signed with warning signs to prevent unauthorized access. A proper monitoring system (hydrogen flow detectors, fire detectors) must be set, adequately maintained, and regularly controlled. All systems must be fail-safe, designed using automatic and passive safety operations when possible (automatic closing valves, automatic alarm system, automatic turning on ventilation if leakage is detected.). Appropriate firefight systems (automatic, when possible) must be implemented. A fire brigade must be in the vicinity, trained specially for hydrogen hazards. Personnel dealing with hydrogen must be well and periodically educated. Protocols and procedures must be well defined. Exposure of personnel must be minimized (both in time and number), proper protective equipment must be used, and safe operational requirements established (e.g walking in pairs).

## 5 Concluding Remarks

Today, many highly innovative companies promise that we will soon be able to fly as regular passengers on hydrogen. The recent realisation that aviation has a huge environmental impact due to use of fossil fuels may have given hydrogen the right momentum to accelerate further research and driving technology validation. As hydrogen-powered aviation proves to be technologically feasible, there has been much discussions about how and when hydrogen-powered aircraft might be used for commercial purposes. Most researchers and experts believe that commercial hydrogen-powered aircraft could enter the market sometime between 2035 and 2050, depending on the type of hydrogen technology (e.g. fuel cell or hydrogen-ICE aircraft) and, of course, the size, with smaller aircraft entering the market earlier than larger ones. Despite several studies and assumptions about the future development of hydrogen-powered aviation, there are still many uncertainties, such as the market applicability of the new technology, the prices of different aviation fuels (e.g. kerosene, biofuel, hydrogen), the capital costs of new aircraft, sufficient renewable resources for hydrogen production, etc. Therefore, only the future can show whether the assumed positive environmental impact of hydrogen-powered aviation will actually materialize.

While hydrogen-fuelled aviation has excellent potential to reduce greenhouse gases, the aspects of the necessary adaptation of hydrogen logistics, airport infrastructure, and hydrogen production, which is essential for the operation of hydrogen-fuelled aviation, also need to be examined. In this context, it is necessary to consider the entire life cycle of hydrogen production since hydrogen technology can only be environmentally friendly if it is produced in an environmentally friendly way (e.g. from renewable resources that do not emit $CO_2$). To enable hydrogen propulsion in aviation. (green) hydrogen production, hydrogen logistics, and airport infrastructure must also be developed. The hydrogen logistics to the airport provides solutions for obtaining, producing, distributing, or transporting hydrogen while also considering storing, purification, liquefaction, or compression of hydrogen at the airport. When considering the logistics of supplying airports with hydrogen, the solutions depends on the size of the airport and the volume of air traffic. At large hubs that will supply many hydrogen-powered aircraft on-site production is likely, while smaller airports will most likely have hydrogen delivered from production sites. For small airports requiring liquid hydrogen, the most feasible solution would be to buy liquid hydrogen from hydrogen producers and deliver it to the airport by $LH_2$ trailer. Airports further away from hydrogen producers may suffer disadvantages due to longer delivery times and costs. If the hydrogen is delivered in gaseous form, liquefaction plants and liquid storage should be placed at the airport, which is less complicated in the case of smaller airports as large airports are subject to face greater spatial restrictions. Nevertheless, this solution is only feasible if more than 20 tonnes of hydrogen are liquefied per day, which is not the case for most airports in Europe that operate flights with up to 70 passengers. If the airport needs a smaller amount of hydrogen, one truck can serve several airports in one delivery trip. Another option is that aircrafts are designed to perform several flights with one tank and get refuelled at the base airport. The airport infrastructure may be changed due to new designs of hydrogen aircraft, which may result in heavier and longer aircraft, impacting the design of runways, taxiways, terminals, gates, passenger bridges, etc. Refuelling infrastructure will need to be adjusted to provide acceptable turnaround rates and adequately manage safety risks by reorganising airport fire locations and equipment. Because smaller hydrogen-powered aircraft used at small and medium size airports are likely to have only minor design changes (e.g., wingspan and weight), such aircraft could be served within existing runways and taxiways without requiring significant adjustments to terminals and stands. At the same time, the refuelling could be organized directly from the truck, causing only a minor increase in traffic at the airport. The smaller aircraft and airports provides an excellent opportunity for the initial applications and testing of hydrogen-powered aviation. Because smaller airports operate fewer flights and usually with smaller aircraft (e.g. fewer than 100 passengers), they are less congested than larger airports or hubs.

Additionally, smaller airports face fewer spatial constraints than hubs, providing more opportunities for necessary expansion or adoption of new infrastructure.

While there are minor technical and safety concerns about the feasibility of hydrogen-powered aviation, significant concern is about the economic and political aspects, i.e. whether policymakers will be able to support 'greener' aviation through appropriate mechanisms (e.g. carbon taxes, emissions trading scenarios, funding research, subsidies), and whether the expected return of an investment will the justify intensive infrastructure investments that will undoubtedly be required to operate hydrogen-powered aviation.

**Acknowledgments** The research was carried out within the MAHEPA project, funded under the European Union's Horizon 2020 research and innovation programme under grant agreement No. 723368. The chapter reflects only the author's view, and the European Union is not liable for any use that may be made of the information contained therein.

## References

European Commission, "An Aviation Strategy for Europe," 2015. [Online]. Available: https://ec.europa.eu/transport/modes/air/aviation-strategy_en. [Accessed: 26-Oct-2020].

E. (NLR) van der Sman, B. (NLR) Peerlings, J. (NLR) Kos, R. (SEO) Lieshout, and T. (SEO) Boonekamp, "Destination 2050-A route to net zero European aviation Preface," 2021.

Air Transport Action Group, "Waypoint 2050: An Air Transport Action Group Project," 2020.

IATA, Aircraft Technology Roadmap to 2050. 2019.

D. Adolf, J.; Balzer, C.; Louis, J.; Schabla, U.; Fischedick; M.; Arnold, K.; Pastowsk, A.; Schüwer, "Shell hydrogen study. Energy of the future? Sustainable Mobility through Fuel Cells and H2," 2017.

McKinsey & Company, Hydrogen-powered aviation A fact-based study of hydrogen technology, economics, and climate impact by 2050, no. May. Belgium: The Print Agency, 2020.

A. P. Brdnik, R. Kamnik, M. Marksel, and S. Božičnik, "Market and Technological Perspectives for the New Generation of Regional Passenger Aircraft," *Energies*, vol. 12, no. 10, 2019, doi: https://doi.org/10.3390/en12101864.

European Commission, "Liquid hydrogen fuelled aircraft—system analysis (CryoPlane), final report," Brussels, Belgium, 2003.

M. Janic, "Is liquid hydrogen a solution for mitigating air pollution by airports?," *Int. J. Hydrogen Energy*, vol. 35, no. 5, pp. 2190–2202, 2010, doi: https://doi.org/10.1016/j.ijhydene.2009.12.022.

D. Verstraete, "On the energy efficiency of hydrogen-fuelled transport aircraft," *Int. J. Hydrogen Energy*, vol. 40, no. 23, pp. 7388–7394, 2015, doi: https://doi.org/10.1016/j.ijhydene.2015.04.055.

M. Janic, "The potential of liquid hydrogen for the future 'carbon-neutral air transport system,'" *Transp. Res. D*, vol. 13, no. 8, pp. 428–435, 2008.

A. Baroutaji, T. Wilberforce, M. Ramadan, and A. G. Olabi, "Comprehensive investigation on hydrogen and fuel cell technology in the aviation and aerospace sectors," *Renew. Sustain. Energy Rev.*, vol. 106, pp. 31–40, May 2019, doi: https://doi.org/10.1016/J.RSER.2019.02.022.

L. Trainelli, C. E. D. Riboldi, A. Rolando, and F. Salucci, "D9.2: Study on hybrid electric powertrain technology and component scalability," 2021.

National Aeronautics and Space Administration, "Liquid hydrogen as a propulsion fuel 1945–1959. NASA-SP-4404," 1978.

B. Khandelwal, A. Karakurt, P. R. Sekaran, V. Sethi, and R. Singh, "Hydrogen powered aircraft: The future of air transport," *Prog. Aerosp. Sci.*, vol. 60, pp. 45–59, 2013, doi: https://doi.org/10.1016/j.paerosci.2012.12.002.

Westenberger; A., "Liquid hydrogen fuelled aircraft—system analysis". Final technical report (publishable version). Cryoplane project, 2003.

Ballard, "Fuel cell and hybrid power systems offer compelling value for UAVs whose missions demand greater runtime than batteries can support," 2018.

Boeing, "Boeing fuel cell plane in manned aviation first," *Fuel Cells Bull*, vol. 4, p. 1, 2008.

Wikipedia, "Boeing Phantom Eye," 2021. [Online]. Available: https://en.wikipedia.org/wiki/Boeing_Phantom_Eye.

DLR, "Antares DLR-H2—out of operation," 2021. [Online]. Available: https://www.dlr.de/content/en/articles/aeronautics/research-fleet-infrastructure/dlr-research-aircraft/antares-dlr-h2-out-of-operation.html.

DLR, "First flight of DLR's HY4 fuel cell light aircraft," *Fuel Cells Bull*, vol. 10, p. 1, 2016.

D. Bachmann, "Dutch Students Just Unveiled the World's First Hydrogen-Powered Aircraft," *Robb Report*, 2021.

FCHEA, "Aviation," 2020. [Online]. Available: http://www.fchea.org/in-transition/2019/11/25/aviation. [Accessed: 28-July-2020].

Y. Ruf, M. Kaufmann, S. Lange, H. Felix, E. Annika, and J. Pfister, *Fuel Cells and Hydrogen Applications for Regions and Cities Vol. 2—Cost analysis and high-level business case*, vol. 2, no. September. GmbH, Roland Berger, 2017.

N. Van Zon, "Liquid Hydrogen Powered Commercial Aircraft," 2012.

N. Sirosh, A. Abele, and A. Niedzwiecki, "Hydrogen Composite Tank Program," *Proc. 2002 US DOE Hydrog. Progr.*, vol. 2000, pp. 1–7, 2002.

D. Verstraete, P. Hendrick, P. Pilidis, and K. Ramsden, "Hydrogen fuel tanks for subsonic transport aircraft," *Int. J. Hydrogen Energy*, vol. 35, no. 20, pp. 11085–11098, 2010, https://doi.org/10.1016/j.ijhydene.2010.06.060.

A. L. Marksel, M., Brdink, A. P., Kamnik, R., Trainelli, L., Riboldi, C. E., & Rolando, "D10.1: Ground infrastructure investment plan," Maribor, 2019.

G. Brewer, "Hydrogen usage in air transportation," *Int. J. Hydrogen Energy*, vol. 3, no. 2, pp. 217–219, 1978.

h2bulletin, "Hydrogen colours codes," 2020. [Online]. Available: https://www.h2bulletin.com/knowledge/hydrogen-colours-codes/

B. Sørensen and G. Spazzafumo, *Hydrogen and fuell cells*, Thrd editi. Joe Hayton, 2018.

G. Naujokaitytė, "Future of green hydrogen is up in the air as the EU dithers over strategy," *Science Business*, Apr-2021.

C. Yang and J. Ogden, "Determining the lowest cost hydrogen delivery mode," *Int. J. Hydrogen Energy*, vol. 32, pp. 268–286, 2007.

IRENA, "Green hydrogen cost reduction," 2020.

S. Gomersall, "Hydrogen in Scotland: The Role of Acorn Hydrogen in Enabling Net Zero," 2021. [Online]. Available: https://pale-blu.com/2020/09/04/hydrogen-in-scotland-the-role-of-acorn-hydrogen-in-enabling-net-zero/

ISO, "ISO/PAS 15594:2004 Airport hydrogen fuelling facility operations," 2004. [Online]. Available: https://www.iso.org/standard/28327.html

C. R. Timmerhaus, K.D.; Mendelssohn, *Advances in Cryogenic Engineering*, Vol 1. New York: Springer, 2007.

Linde, "Hydrogen," 2021. [Online]. Available: https://www.linde-engineering.com/en/process_plants/cryogenic_plants/hydrogen_liquefiers/index.html

ISO, "ISO/TR 15916:2015(en) Basic considerations for the safety of hydrogen systems," 2015. [Online]. Available: https://www.iso.org/obp/ui/#iso:std:iso:tr:15916:ed-2:v1:en

Hydrogen Europe, "Aviation," 2020. [Online]. Available: https://hydrogeneurope.eu/aviation-0. Accessed 28 July 2020

B. Marshall, "How the Hydrogen Economy Works," 2021. [Online]. Available: https://auto.howstuffworks.com/fuel-efficiency/fuel-economy/hydrogen-economy.htm.

W. A. Amos, "Costs of Storing and Transporting Hydrogen," 1998.

NASA, "LH2 Airport Requirements Study, CR2700," Washington, D.C., 1976.

A. Brinchmann, "Hydrogen as seen from a shipping company," *H2@Ports Workshop San Francisco*, 2019.

J. Saul, "The Race is On to Pioneer Hydrogen Shipping," 2021.

L. Blain, "Kawasaki launches the world's first liquid hydrogen transport ship," *New Atlas*, 15-Dec-2019.

"KSOE and Korean Register to develop hydrogen vessel standard," *Ship Technology*, 2021.

ACI/ATI, "Integration of Hydrogen Aircraft into the Air Transport System," 2021.

NFPA, "NFPA 2: Hydrogen Technologies Code," 2021. [Online]. Available: https://www.nfpa.org/.

E. Fleuti, "Aircraft Ground Handling emissions at Zurich airport—methodology and emission factors," 2014.

ICAO, "ICAO (2010) Operational Opportunities to Minimise Fuel Use and Reduce Emissions," Montreal, Quebec, Canada, 2010.

W. Burmeister, "Hydrogen project at the Munich Airport," 2000.

E. Testa, C. Giammusso, M. Bruno, and P. Maggiore, "Analysis of environmental benefits resulting from use of hydrogen technology in handling operations at airports," *Clean Technol. Environ. Policy*, vol. 16, no. 5, pp. 875–890, 2014, https://doi.org/10.1007/s10098-013-0678-3.

I. May, "Green hydrogen facility opens at Berlin airport," *Fuel Cells Bull.*, vol. 2014, no. 5, p. 1, 2014, https://doi.org/10.1016/s1464-2859(14)70122-1.

Air Liquide, "A new step forward in clean mobility: Air Liquide and Groupe ADP open the hydrogen station at the Paris-Orly Airport," 2017. [Online]. Available: https://energies.airliquide.com/new-step-forward-clean-mobility-air-liquide-and-groupe-adp-open-hydrogen-station-paris-orly-airport. [Accessed: 18 Aug 2020].

Fleet Europe, "On-site hydrogen for Liege Airport vehicles," 2018. [Online]. Available: https://www.fleeteurope.com/en/smart-mobility/belgium/article/site-hydrogen-liege-airport-vehicles?a=THA13&t%5B0%5D=CMI&t%5B1%5D=LiegeAirport&curl=1. [Accessed: 18-Aug-2020].

FuellCellWorks, "Bremerhaven airport becomes location of hydrogen pilot plant," 2020. [Online]. Available: https://fuelcellsworks.com/news/bremerhaven-airport-becomes-location-for-hydrogen-pilot-plant/. [Accessed: 17-Aug-2020].

Air Liquide, "Air Liquide is partnering with Incheon Airport, Hyundai Motor Company and HyNet to deploy hydrogen stations at Seoul Airport," 2020. [Online]. Available: https://energies.airliquide.com/air-liquide-partnering-incheon-airport-hyundai-motor-company-and-hynet-deploy-hydrogen-stations. [Accessed: 18-Aug-2020].

C. E. D. Riboldi, L. Trainelli, and A. Rolando, "MAHEPA D10. 2: Cost models for hybrid electric aircraft," 2021.

Our airports, "Open data," 2020. [Online]. Available: https://ourairports.com/data/. [Accessed: 26-Oct-2020].

C. Amy and A. Kunycky, "Hydrogen as a Renewable Energy Carrier for Commercial Aircraft," 2019.

U. Schmidtchen, E. Behrend, H.-W. Pohl, and N. Rostek, "Hydrogen aircraft and airport safety," *Renew. Sustain. Energy Rev.*, vol. 1, no. 4, pp. 239–269, Dec. 1997, https://doi.org/10.1016/S1364-0321(97)00007-5.

Chose Paris region, "Turning the Airport into a Hydrogen Ecosystem," 2021. [Online]. Available: https://www.h2hubairport.com

ISO, "ISO standards," 2021. [Online]. Available: https://www.iso.org/advanced-search/x/title/hydrogen/status/P/docNumber/docPartNo/docType/0/langCode/ics/currentStage/true/searchAbstract/true/stage/stageDateStart/stageDateEnd/committee/sdg.

Wikipedia, "Hindenburg disaster," 2021. [Online]. Available: https://en.wikipedia.org/wiki/Hindenburg_disaster.

# Fuel Cells as APU in Aircrafts

**Samuel Tadeu de Paula Andrade, Marina Domingues Fernandes, Victor N. Bistritzki, Rosana Zacarias Domingues, and Tulio Matencio**

## 1 Introduction

For the next 30 years, we will face a drastic transformation in the way we transport. Climate change poses significant challenges for the heavy-load and long-distance transportation segment, including busses, trucks, ships, and aircraft. Existing fossil fuels have to be cut by 45% until 2030 and net-zero by 2050, as governments agreed in the Paris Convention [1]. Hydrogen and its related technologies are a central pillar to achieving the Convention's goal of limiting global warming to1.5 °C. They are a growing interest of major aviation companies and researchers [2]. Hydrogen is a potential free-emission fuel that can enrich conventional jet fuels to produce carbon-neutral synthetic fuels or directly combusted. An electrochemical fuel cell device can directly convert hydrogen to electricity at a higher efficiency than traditional combustion processes. The electricity can power the aircraft propulsion system or the aircraft auxiliary power unit (APU).

The use of fuel cells in aviation is a realistic alternative that aviation companies and manufacturers consider. Thus, fuel cells are included in short-, medium-, and long-term research and development that aim at reducing $CO_2$ emissions in the aviation industry. The feasibility of fuel cells depends on a reliable infrastructure for hydrogen or other alternative fuels. In addition, drop-in electrofuels, synthetic fuels, and other fuel technologies have been steadily investigated and tested with significant financial support.

S. T. de Paula Andrade · M. D. Fernandes · V. N. Bistritzki · R. Z. Domingues
T. Matencio (✉)
Universidade Federal de Minas Gerais, Belo Horizonte, Minas Gerais, Brazil
e-mail: samuelt@ufmg.br; marinadf@ufmg.br; bistritzki@ufmg.br; rosanazd@ufmg.br; tmatencio@ufmg.br

 
C. O. Colpan, A. Kovač (eds.), *Fuel Cell and Hydrogen Technologies in Aviation*, Sustainable Aviation, https://doi.org/10.1007/978-3-030-99018-3_7

At some airports, like in Memphis and London Heathrow, fuel cells powered by hydrogen or other available fuels are already implemented as backup power generators. The hydrogen infrastructure development at airports is a strategic entry point for this technology to the aviation sector. Safety standards and regulations of operating conditions can be validated, and hydrogen can be safely integrated into airport boundaries. When introducing fuel cells to aircraft, the most immediate application is utilizing them as APUs [3]. Fuel cell electric power can be used for ground or on board operations.

A significant advantage of a fuel cell APU on board is that it can be integrated into most conventional aircraft architectures without compromising fuselages. This compatibility facilitates integration and avoids the high costs of redesigning aircraft architectures or developing new aircraft in the initial testing phases. Taxiing is an example where fuel cell APUs can improve air quality. The aircraft operates at low speed, low fuel efficiency, and high $CO_2$ emissions during this maneuver. Electric taxiing systems, operated by fuel cells, address this issue as combustion of jet and fossil fuels are avoided [4].

The fuel cell is an electrochemical device that converts the chemical energy of a fuel to electrical energy. Fuel cells with a technology readiness level (TRL) higher than 6 are suitable for supplying energy for APUs. Those are Proton-Exchange Membrane Fuel Cell (PEMFC) [5], Solid Oxide Fuel Cells (SOFC) [6], and Direct Methanol Fuel Cells (DMFC) [7]. This chapter will primarily focus on PEMFCs and SOFCs—the most viable technologies for APUs in aircraft.

In these three fuel cell types, the fuel enters on the anode side (negative electrode), where part of the electrochemical reactions occur. An electron flow passes through an external circuit (load) and then reacts on the cathode (positive electrode), which is fed with oxygen. The subproduct of these reactions is mainly water.

SOFCs are based on electrochemical reactions that use the oxygen ions as a carrier. Usually, the electrolyte is made of stabilized yttria zirconia (YSZ), the anode is a compound of YSZ and Ni, and the cathode is based on strontium-doped lanthanum manganite (LSM). SOFCs operate at temperatures between 650 and 800 °C, allowing oxidation and reduction reactions to occur without rare metal catalysts such as platinum. They are also resistant to sulfur and carbon monoxide poisoning.

PEMFCs are based on electrochemical reactions that use hydrogen ions as a carrier through an electrolyte. This electrolyte is made of perfluorosulfonic acid (PFSA) polymer. The operating temperature is low, around 80 °C, which enables fast warm-up. However, the anode and cathode of the PEMFC are made from expensive platinum-based materials.

DMFCs are fed by pure methanol, which is usually mixed with water and fed directly to the fuel cell anode. Usually, the DMFC electrolyte, anode, and cathode are made from similar materials as a PEMFC. However, DMFC can suffer from methanol crossover (when it crosses the electrolyte without fully reacting at the anode) and self-poisoning. Figure 1 visualizes the operating scheme of the three types of fuel cells.

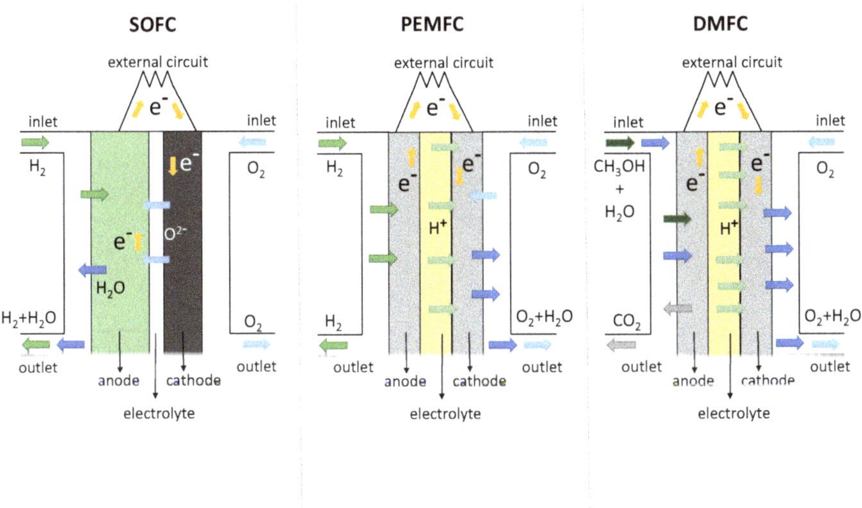


**Fig. 1** Operating scheme of SOFC, PEMFC, and DMFC. (Source: the authors)

# 2 APU

## *2.1 Conventional APU*

Usually, an aircraft is equipped with the main engine that generates thrust and an auxiliary power unit designed to partially supply the aircraft's electrical, pneumatic, and hydraulic loads. The APU system is powered by the same fuel as the aircraft and is composed of a compressor, turbine, burner, shaft, and an electric generator, as shown in Fig. 2a. Despite being powered by the same aircraft fuel, the APU system is independent of the main engine. Therefore, the APU system provides electrical energy for operating several aircraft components but does not generate thrust. This system operates based on the open Brayton cycle coupled with an electrical generator. The Brayton cycle is composed of isentropic compression and expansion processes, in addition to two isobaric heat exchanges: one of which occurs in the burner and the other in the atmosphere, as shown in Fig. 2b.

The APU provides both shaft and pneumatic power. The former is used to move the electric generator and can also move hydraulic pumps or any other system that requires energy. The latter is used to start the main engines, power the de-icing and pressurization system, and the air conditioning.

The use of conventional APUs in aircraft began in the 1900s, and by the 1960s, they were widely implemented in commercial aviation. APUs enable ground autonomy, which allows the aircraft to be energy independent from the airport structure and ground power units (GPUs). An APU system can operate for a few minutes or even several hours, and it is mainly used for ground operations like taxiing, embarking, and disembarking. The system can also be used in unusual in-flight situations

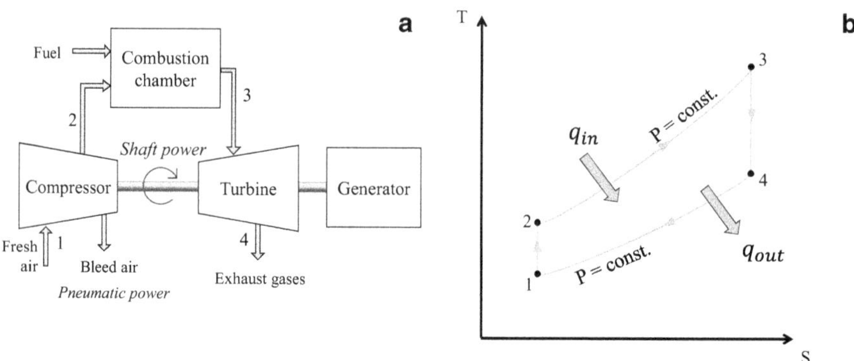


**Fig. 2** APU simplified Scheme (**a**) and Brayton cycle with the isentropic compression and expansion processes (**b**). (Source: the authors)

such as restarting the main engines and emergencies. In the case of failure of the main generators and the APU, the ram air turbine (RAT) provides electricity for the aircraft. The APU is localized in the tail cone on most aircraft, and the exhaust gas is released directly into the atmosphere.

Several manufacturers design and provide commercial APU systems. The technology of those systems is essentially the same, although the rating power varies according to the aircraft [8]. Table 1 presents some APU models' mass and rating power values [9].

Pneumatic power produced can range from 7 kW, with a system weight of 27 kg for small commercial aircraft, up to 48 kW, and 263 kg for large commercial aircraft, such as a Boeing 747 [10, 11]. The APU operates at a constant frequency of 400 Hz and with voltages between 115 and 230 Vac. The APU's electrical efficiency is relatively low, around 15–20%, due to the reduction in air density with the increase in attitude. The low efficiency is one of the reasons to use generators integrated into the main engines, which generate electric energy with efficiencies around 50% [11, 12].

Despite their notorious disadvantages (e.g., noise production and pollutant emissions), APUs are relatively light and use the same fuel as the one used for the main engine, which is typically jet fuel. However, several policies are being applied to airports [12–14]. One of these policies includes the taxation of airlines based on the level of noise produced at Heathrow, Gatwick, and Stansted airports. Similar policies in terms of pollutant emissions are observed in the Zurich airport. The International Civil Aviation Organization proposed in 2011 an agreement to reduce fuel consumption at airports by up to 2% between 2020 and 2050 [15]. APUs are emitting 19% of all nitrogen oxides ($NO_x$) at London Heathrow, and according to estimates of 325 US airports, APUs are responsible for emitting 10–15% of carbon monoxide (CO) and 15–30% of sulfur oxides ($SO_x$) and $NO_x$ [14].

**Table 1** Characteristics of some commercial APUs

| APU manufacturer | Model | Mass (kg) | Rating power (kW) | Power density (kW/kg) |
|---|---|---|---|---|
| Aerosila | TA-6A | 245 | 235 | 0.959 |
| Aerosila | TA-12 | 290 | 287 | 0.990 |
| Saphir | Saphir 100 | 78 | 135 | 1.731 |
| Honeywell | 36-300 | 136 | 291 | 2.140 |
| Honeywell | 131-9A | 159 | 343 | 2.157 |
| Honeywell | RE220 | 109 | 261 | 2.394 |
| Honeywell | 36-100 | 53 | 127 | 2.396 |
| Honeywell | RE100 | 40 | 101 | 2.525 |
| Honeywell | 36-150CF | 53 | 142 | 2.679 |
| Sundstrand | APS2000 | 132 | 376 | 2.848 |
| PrattWhitney | PW901 | 379 | 1145 | 3.021 |
| Sundstrand | APS3200 | 138 | 450 | 3.261 |

Source: the authors

With the advancement of the More Electric Aircraft (MEA)[1] concept, the demand for more significant amounts of electric energy has increased expressively. In the MEA concept, systems that were once mechanical, pneumatic, and hydraulic are replaced by electrical systems in search of higher efficiency, reliability, and reduction of costs and pollutant emissions [11]. Thus, MEAs favor the entrance of APUs powered by clean fuels or even the use of fuel cells.

## 2.2 *Aircraft Fuel Cell APU*

The environmental issues and demand for higher electrical power open opportunities for the use of fuel cells as an APU due to their potential efficiencies between 50 and 70% [10, 15, 17–20].

Despite the quality of electrical energy in MEA, the higher efficiency of fuel cell APU can be improved as they can be placed close to where the energy is demanded, like the wings, not just in the tail cone. In MEA, starting the main engine is no longer pneumatic and becomes electrical, with an additional electrical charge supplied by the APU [11, 21]. Figure 3 compares the electrical power generated on main engines with the maximum takeoff weight (MTOW) of conventional aircraft and MEA [9, 22, 23].

In aerospace applications, the first fuel cells were PEMFCs and alkaline fuel cells (AFC) in the Gemini and Apollo space programs [24, 25]. The Soviet N1-L3

[1] The MEA is a broad concept which focuses on replacing as many supporting aircraft systems based on pneumatic, hydraulic, and mechanical power as possible, with systems based on electric power. The advantages of this concept are improved reliability, easier maintenance, and greater efficiency [16].

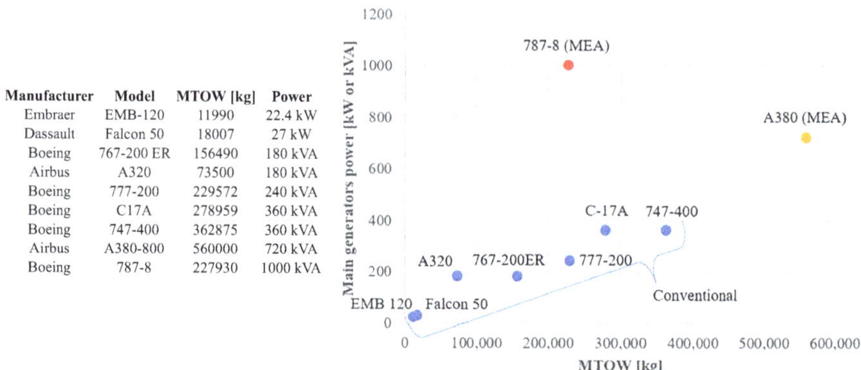

| Manufacturer | Model | MTOW [kg] | Power |
|---|---|---|---|
| Embraer | EMB-120 | 11990 | 22.4 kW |
| Dassault | Falcon 50 | 18007 | 27 kW |
| Boeing | 767-200 ER | 156490 | 180 kVA |
| Airbus | A320 | 73500 | 180 kVA |
| Boeing | 777-200 | 229572 | 240 kVA |
| Boeing | C17A | 278959 | 360 kVA |
| Boeing | 747-400 | 362875 | 360 kVA |
| Airbus | A380-800 | 560000 | 720 kVA |
| Boeing | 787-8 | 227930 | 1000 kVA |



**Fig. 3** Electric power generated by the main engines and maximum takeoff weight (MTOW). (Source: the authors)

space program also used fuel cells for the lunar module [26]. More than in other transportation types, fuel cell application in aircraft requires optimized mass, standard rigid safety criteria, and high reliability. Some studies indicate the following desired characteristics: power densities between 0.75 and 1.0 kW/L, specific power of 0.5 kW/kg to more than 1 kW/kg, and durability between 20,000 and 40,000 h [19, 21, 27].

Besides producing energy, fuel cell APUs can:

- Provide water for consumption in toilets and galleys.
- Heat water.
- Power the wing de-icing system.
- Provide inertizing gas for the On Board Inert Gas Generation System.
- Replace the RAT.
- Provide heat to the Environmental Control System.
- Power the electric motors on the MEA wheels during taxiing.

For example, a PEMFC can produce about 0.5 L of water for every kWh of electrical energy produced. Therefore, it is possible to yield 975 L of water, enough to supply a Boeing 787, with a tank with a maximum capacity of 1022 L [13, 15, 28–30].

Recent studies indicate SOFC and PEMFC as the most viable options for aeronautical applications. Both can be used for propulsion or APU purposes [12, 20] and have TRL higher than 6. Figure 4 illustrates a general and simplified diagram of a fuel cell APU system, applicable for SOFC and PEMFC and presents the main subsystems, including the typical inputs and outputs of a fuel cell APU. The level of complexity depends on the type of fuel cell stack and the expected functions of the system. The Balance of Plant (BoP) generally comprises equipment such as compressors, pumps, turbines, vessels, humidifiers, reformers, heat exchangers, burners, and other occasional components. Stable powered electrical loads are powered by the fuel cell, while the battery powers dynamic loads. With the support of sensors and actuators, the controller appropriately manages all these items [20, 31].

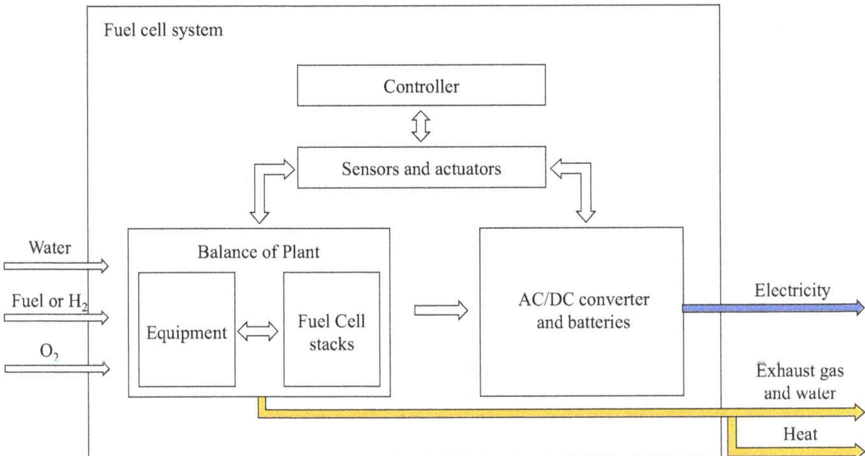


**Fig. 4** Simplified diagram of a fuel cell APU applicable for SOFC and PEMFC. (Source: The authors)

The gas oxygen, $O_2$, consumed on the cathode side of a fuel cell can come from three sources. First, from the outdoors, which produces extra drag. Second, from the bleeding of the main engines, which interferes with their functioning. And third, from the passenger cabin, which may not meet the required flow rate for the fuel cells or the cabin safety and control criteria. Scientific studies investigate these different options and their challenges, especially when using outside air. With no other compression system, the increasing altitude reduces air density and consequently decreases the oxygen pressure on the cathode side, which is harmful to the fuel cell performance [12, 21, 27, 32, 33].

In general, PEMFCs find their main application in smaller aircraft with lower electrical power consumption. PEMFCs demand high hydrogen purity, which requires that the hydrogen is stored on board the aircraft. The consequence is that the larger the aircraft, the larger the electrical power consumption; therefore, the larger the hydrogen storage solution must be. The storage system becomes cumbersome and loses competitiveness to the conventional APU in larger aircraft. Using PEMFCs in larger aircraft makes it possible to employ reformers with other purification systems; however, like larger tanks, they also add mass to the entire system. An advantage of PEMFCs over other fuel cell types is their operation temperatures of around 80 °C, their high specific power ratio of up to 1.2 kW/kg, and their low start-up time. Furthermore, PEMFC technologies are, to date, further developed than other fuel cell types, mainly due to their advances in mobile and automotive applications [12, 20, 29, 32, 34, 35].

Another challenge of PEMFC-APUs is water management. The ohmic loss[2] of the ion exchange membrane increases with the humidity, influencing the cell's temperature. However, if carefully managed, these variables, for example, when operating with the hydrogen circuit in the dead-end configuration, can lead to high utilization factors [36–38].

SOFC technology is more suitable for larger missions and aircraft with large energy consumption. The high operating temperature of around 800 °C enables the integration of a cogeneration turbine, which leads to higher system efficiency. In this case, the fuel cell tends to operate pressurized and, therefore, a trade-off study among pressure, system mass, and efficiency should be carried out. In addition, SOFCs can operate with internal or external reforming from the jet fuel as they are more tolerant to impurities than PEMFC. Jet fuel, however, can contain sulfur, and as SOFCs require low levels of sulfur, the use of desulfurizers in fuel cell systems may be necessary. A significant advantage of the SOFC over PEMFC is that they do not rely on hydrogen storage solutions, which reduces the entire system mass despite the possible use of a reformer and/or desulfurizer. However, due to the low ionic conductivity of ceramic electrolytes, SOFC operation demands high temperatures, which affect sealing systems and long start-up times [12, 21, 27, 29, 33, 39–41].

Using fuel cell APUs reduces pollutant emissions significantly compared to conventional APUs as zero or negligible emissions of $NO_x$, CO, and unburned hydrocarbons are possible [21, 27, 36]. The Society of Automotive Engineers International (SAE) has been implementing standards, guides, and technical guidance for fuel cells for aircraft application. Valid for PEMFC and SOFC, the technical report AIR1408 [42] contains an extensive list of standards with system requirements for transferring fuels to fuel cells, and the technical report AIR7765 [43] presents an overview of the use of fuel cells in aircraft. Precisely, for PEMFC, two documents guide the use of this fuel cell type in aircraft. The first is the technical specification AS6858 [31] that elaborates on the requirements for designing, testing, and certifying fuel cells. The second is the technical report AIR6464 [44], with the main safety requirements for integrating fuel cells into aircraft.

R&D reports that mentioned the feasibilities of fuel cell APUs are available in the literature, but main issues and challenges remain. The total mass of fuel cell APUs can be higher than that of a conventional APU. This issue reduces payload capacity and requires optimization to ensure performance while guaranteeing adequate financial returns [12]. As previously mentioned, the fuel to operate a PEMFC as APU cannot contain any contaminants such as sulfur and CO. There are two solutions to this problem: (1) fuel reforming or (2) hydrogen storage in the aircraft. However, the first option comes with a higher mass due to a greater quantity of components. The second option leads to logistical problems at airports and an increase in weight and volume on the aircraft compared to the conventional storage tank [12]. As the energy content of hydrogen per volume and density is low, the

[2] The ohmic losses are related to Joule heating, which is the conversion of electric power to thermal power. This process occurs by the electron's interactions with the matter's atoms.

hydrogen storage requires a pressure of up to 700 bar and/or cryogenic vessels [29, 43].

As SOFCs incorporate ceramic material, the effects of vibration and mechanical impacts on those more fragile materials are more critical and must be analyzed. The increase of the specific power and reduction of the mass/volume ratio of the desulfurizer and reformer should also be investigated further. Furthermore, the high operating temperatures cause issues of start-up time and temperature transients. Indeed, the greater the level of integration of the APU functions that would entirely or partially replace conventional systems, the more significant the possible weight reduction [12, 21]. For both SOFC and PEMFC, production scale studies and their respective cost reduction possibilities must be considered so that fuel cells can be cost-competitive concerning conventional APUs [12, 21, 45]. The following Table 2 presents a comparison between the conventional and fuel cell APU characteristics.

# 3 Fuel Cell APU Technological Developments

## *3.1 Aircraft PEMFC-APU*

The first crewed flights in the history of civil aviation using a PEMFC-APU system were carried out in 2007 and 2008 by Airbus in collaboration with Michelin and the German Aerospace Center DLR [47]. During a standard mission, flights of an A320 aircraft, at an altitude of 25,000 ft. and equipped with the hybrid 20 kW APU system, the hybrid PEMFC-APU was evaluated using hydrogen as fuel and oxygen as an oxidizer in different acceleration and lean conditions. The system's behavior proved to be satisfactory under the flight conditions that considered, besides other parameters, pressure, vibration, temperature, and altitude. In addition to the PEMFC's electric power generation function, satisfactory production of water (0.5–0.6 L/kWh), which could be used inside the aircraft (toilet, air conditioning, etc.), and oxygen-poor exhaust gases (only 10% V/V of $O_2$), adapted for fuel tank safety (retardant fire and suppressing explosions) were verified.

An advantage of the hybrid PEMFC-APU system is its use at airports to electric power motors on the aircraft wheels, thus allowing taxiing during ground operations. With the PEMFC-APU multifunctional approach, energy efficiency on board is optimized. Most projects in the area [13, 19, 28] conceptualize and develop innovative electrical architectures. The PEMFC-APU's multifunctional system is a central component for efficient future aviation.

Due to the technical and financial complexity of experimental fuel cell APU tests, most experimental research is conducted on smaller aircraft and laboratories equipped with test platforms that simulate aerospace navigation operations [28, 30, 32, 36, 47–51]. Other studies focus on developing mathematical models, generally validated from laboratory experiments, that optimize and integrate various fuel cell APU components (like compressor, turbine, PEMFC, reformer, etc.) [13, 15, 19, 20,

**Table 2** Summary for comparison of traditional APU and FC-APU

| Characteristics | Traditional APU | SOFC/PEMFC APU | Commentaries |
|---|---|---|---|
| Efficiency | 20–30% | 40–50% (PEMFC) 40–70% (SOFC) | Significant energy gain using fuel cell APU |
| Fuel | Kerosene (jet fuel) | Hydrogen (PEMFC, SOFC) Internal hydrocarbon reforming (SOFC) | A storage system for hydrogen is needed when it is used as a fuel for the fuel cell |
| Starting the jet engines | Yes | Yes | There is always a start-up time with fuel cell APU, especially SOFC-APU, due to the need to heat the system |
| Taxiing | Yes | Yes | Decrease of pollutant emissions, external noise, and wear of jet engines when using fuel cell taxi |
| APU location | Generally, in the tail cone of the aircraft | It can be distributed on aircraft | In fuel cell APUs, energy losses are reduced by reducing the distance between the energy source and electrical loads |
| Water | Stored in aircraft tanks | Produced by the APU | Less weight on take-off when using a fuel cell APU. The water released during the operation of the fuel cells can be used in the aircraft |
| Emergency energy | RAT | The fuel cell can substitute the RAT | In the case of a fuel cell APU, a RAT is not needed |
| Security/fuel tanks | Gas separation system | Use of inert gases emitted by the fuel cell | In the case of the APU, there is a need to ship gas for blanketing |
| Weight | Low | High | Increase in weight in the case of the fuel cell APU but can be offset by not having to carry water |
| Noise pollution | High | Low | Passenger comfort and following new airport policies |
| Environmental pollution | High | Low | Under new airport policies |

Adapted from [46]

34, 37, 38, 52]. Some scientific contributions address the specific issue of fueling the hybrid PEMFC-APU system [53–55].

Due to the importance of pressure on the performance of fuel cells and considering the decrease in atmospheric pressure at high altitudes, Pratt and colleagues [32] carried out unprecedented laboratory experiments, which tested the effect of sub-atmospheric pressure on the operating voltage of a PEMFC fueled with pure hydrogen. A stack composed of 23 cells, each having an active area of 13 cm$^2$, was placed in a vacuum chamber that allowed the simulation under pressure conditions associated with altitudes of up to 16,307 m. As expected, as altitude increases, pressure decreases, and the PEMFC strain reduces. The study concluded that the relationship between the decrease in PEMFC performance and the potential losses due to activation, ohmic resistance, and gas concentration is essential to determine the characteristics of PEMFC for application in aircraft.

Horde and colleagues [49] also analyzed the influence of altitude on the functioning of a PEMFC. The group investigated the behavior of PEMFC at three different altitudes (200, 1200, and 2200) that represented a small aircraft flight and tested under different conditions of air stoichiometric factors (from 1.5 to 2.5). The tested system (Bahia Helion PEMFC system) consisted of a stack of 24 cells with 100 cm$^2$ active area each (1.2 kW), which operated with pure hydrogen, a compressor, a heat exchanger, and a humidifier. The results showed a decrease in the PEMFC potential due to the decrease in the pressure on the cathode side. The decrease is expected when considering the effect of partial pressure of oxygen on the theoretical equilibrium potential of the fuel cell, called the Nernst potential, $E_{\text{nernst}}$, and expressed by Eq. (1) [49].

$$E_{\text{nernst}} = \frac{\Delta G}{2F} + \frac{\Delta S}{2F}\left(T_{\text{FC}} - T_{\text{ref}}\right) + \frac{RT_{\text{FC}}}{2F}\left[\ln\left(P_{\text{H}_2}\right) + \frac{1}{2}\ln\left(P_{\text{O}_2}\right)\right] \tag{1}$$

where $\Delta G$ and $\Delta S$ are, respectively, the variations of free energy and entropy of the electrochemical reaction, $F$ is the Faraday constant, $R$ the perfect gas constant, $T_{\text{FC}}$ the unit cell temperature, $T_{\text{ref}}$ the reference temperature (generally 20 °C), and $P_{\text{H}_2}$ and $P_{\text{O}_2}$ the pressures of hydrogen and oxygen, respectively.

The stack potential, $U$, is directly related to the Nernst potential, as can be seen in Eq. (2) [49].

$$U = n_{\text{cell}} \times \left[E_{\text{Nernst}} - U_{\text{act}} - U_{\text{diff}} - U_{\text{Ohm}}\right] \tag{2}$$

where the total stack potential, $U$, is a function of the number of unit cells present in the stack, $n_{\text{cell}}$, the theoretical equilibrium potential of fuel cells, $E_{\text{Nernst}}$, and the losses caused by the passage of current and related to: (1) charge transfers in fuel cell electrodes, $U_{\text{act}}$, (2) mass transport of electroactive species, $U_{\text{diff}}$, and (3) the ohmic drops related to the electrical resistances of unit cells, $U_{\text{ohm}}$.

When the PEMFC is operating in a high-altitude, low-temperature environment, a certain degree of pressurization is required to allow for the presence of liquid water necessary for the proper functioning of the PEMFC. Under these conditions, Campanari and colleagues [37] explored using a turbine to expand the fuel cell exhaust, resulting in a turbocharged fuel cell project. Therefore, in all cases, the ideal pressure level of the system and its best layout must be defined. Thus, the authors carried out an energy feasibility study comprising three possible architectures with a dead-end anode PEMFC and 60 kW electrical power. The fuel cell operated on pure hydrogen in three flight conditions: ground operation and intermediate and high-altitude cruise conditions. The architectures differ mainly in the compression system setup. In the first setup, the compressor is powered by an electric motor. In the second, the compressor uses a turbine to take advantage of the stack's exhaust gases. And in the third, a burner powered by jet fuel is located before the turbine. The second architecture showed the best electrical efficiency reaching a maximum value on the ground with 53.04%. In cruise, the electrical efficiency of 50.53% was much higher than the one of a conventional APU, which was only 17.75%. It is worth noting that PEMFC operates with pure hydrogen and requires adequately humidified air. Usually, increased pressure promotes increased efficiency in architecture, except when the inlet pressure is very low.

The researchers that were responsible for the first in-flight test of the PEMFC-APU hybrid system [47] developed theoretical/experimental research on the production of electricity, water, and inertizing gas from a PEMFC-APU system [56]. After discussing the requirements of a conventional APU system defined by the Federal Aviation Administration (FFA), the authors demonstrated the benefits of systems with pressurized hydrogen storage tanks and cabin air based on a theoretical and experimental simulation project aircraft cruise and ground operation.

Research from Keim and colleagues [28], within the projects "EFFESYS" and "BRIST" financed by the German government, aimed at increasing efficiency and safety and reducing noise and environmental pollution. The group tested a hybrid system with a PEMFC-APU in collaboration with Airbus's Fuel Cell Team. The use of multifunctional systems combines the production of electricity, heat, clean and demineralized water (internal use in the aircraft for toilets, kitchens, and humidification), and inertizing gas from fuel tanks. The PEMFC produced about 0.5 L of water per kWh of electricity produced. Theoretically, as it is possible to produce 975 L of water during a conventional Boeing 787 flight, its water tank (maximum capacity of 1022 L) does not need to fill the tank before takeoff. The possibility of producing water during flight leads to a significant reduction of the starting weight of the aircraft [28, 56]. The PEMFC of the system tested was a Hydrogenics HyPM Xr12, with a maximum power of 12 kW, weighing 80 kg, and having an electrical efficiency of 53%.

The DLR's *Luftfahrtforschungsprogramm* (LuFO) studied a hybrid APU system using a PEMFC HyPM Xr12 to evaluate the theoretical effect of low pressure and the effect of supercharging on aircraft performance [51]. It provided evidence that PEMFC operation under low pressure is technically feasible (700 mbar). The study showed that, by using an external compressor, the energy demand to pressurize the PEMFC is almost the same as the energy loss of the PEMFC due to low-pressure operation. The results indicated that a pressure-dependent self-humidification control could be a solution to improve the performance of PEMFC.

Moreno and colleagues [15] simulated the use of a PEMFC-APU, powered by stored hydrogen, to supply the entire electrical load of an Airbus A320 and two electric motors to perform the taxiing maneuvers. The simulated system demonstrated a maximum capacity of up to 150 kW, mass between 240 and 350 kg, and 1.3 kg of hydrogen consumption to complete an entire taxiing cycle. In comparison, a conventional APU provides up to 90 kW, with a mass of 160 and 6.76 kg of jet fuel consumption.

Peters et al. [19] theoretically evaluated fuel cell technologies for multifunctional use in aircraft. Besides the electrical system efficiency, the authors emphasized water production and the availability of gases for the tank inertization. The group proposed a multistep process analysis methodology to select the most appropriate fuel cell system configuration. A systematic analysis of the process was performed by introducing parameters with the aid of statistical tools. The addressed fuel issues showed that a possible 95–96% $CO_2$ reduction in the APU operation could be achieved by using biofuels derived from wood. The overall evaluation indicates that hydrogen-based systems are more advantageous in achieving high efficiency with high water production rates. Considering the volume and mass balances, PEMFCs are preferable for short-range missions. High-temperature PEMFC systems, based on jet fuel, are a better choice for medium to long-range missions.

Samsun and colleagues [55] studied a high-temperature PEMFC-APU with an autothermic reformer and catalytic burner. The study carried out a proof of concept for an integrated 5 kW high-temperature PEMFC system operating on diesel and jet fuel. The system was tested with GTL (gas-to-liquid) jet fuel, BTL (biomass-to-liquid) diesel, and premium diesel oil. All fuels achieved the electric power target of 5 kW, and the operation of the self-sustaining system was demonstrated. In a 250-h operation, the system showed a slight drop in performance.

Guida and Minutillo [34] developed a thermal model coupled with an optimization method to design an APU for a light aircraft (Aermacchi SF 260). The system has 28 stacks, five air compressors, six hydrogen tanks, six batteries, and one heat exchanger. At an altitude of 3000 m and with a total mass of 282 kg, it can supply 24 kW for 6 h. The tanks account for 47% of the total mass and the stacks for 24%.

Thirkell and colleagues [20] developed a model based on the characteristics of 527 aircraft and divided them into 15 categories. The model was based on the following entries: the aircraft category, propulsion means, fuel cell purpose (propulsion or APU), maximum takeoff weight, autonomy, and altitude. Using this model, the group estimated the PEMFC power output, number of cells, cooling mode, mass, and system volume. For small and light aircraft, such as an unmanned aerial vehicle (UAV), the use of PEMFC proved to be viable. However, the study revealed that PEMFC systems are unfeasible for large aircraft due to excessive weight. For example, a 162 kW PEMFC system necessary to power an Airbus A320 would weigh around 2500 kg, including hydrogen and oxygen storage. For these cases, the author indicates the use of SOFC since it can operate with liquid fuels.

## 3.2 *Aircraft APU-SOFC*

SOFCs and PEMFCs share many common characteristics for their use in aviation compared to conventional APU systems. Both can supply power to electrical loads and water, oxygen, inertization gases, and other advantages, highlighted in the previous section. This section will focus on the specificities of APU-SOFCs.

The use of fuel cells in aircraft has been explored since 1984 [57]; however, only after 2004, SOFCs started to be investigated, especially as a secondary energy source such as APUs [58, 59]. Research on SOFC-APUs is limited to theoretical studies due to the high costs and technical and safety difficulties of conducting in-flight experiments [60]. Reports of experimental tests aimed at application in the stationary energy sector [61–63], land transportation [64, 65], and maritime [66, 67]. Fernandes et al. [68] conducted exhaustive literature and patent reviews on SOFC-APU systems for aircraft. Although experimental studies are still limited, the review presents clear potentials for using SOFC-APUs in aircraft. Table 3 shows different power ranges and applications (regional mission airplanes and fully electric UAVs, short-range mission airplanes, and long-range mission airplanes) studied by modeling SOFC-APU systems.

Conceptual studies of APUs in commercial aircraft reveal many advantages of SOFCs over PEMFCs [84]. Different architecture models explore energy efficiency advantages of SOFC-APUs that are mainly related to the high operating temperature and enable cogeneration by including heat exchangers, mixers, pump reformers, and compressor-turbine [33, 40, 60, 81, 85–88].

A characteristic of SOFCs is that they generate heat due to exothermic electrochemical reactions. When integrated into an APU and operated with a compressor-turbine, the SOFC exhausted heat can be availed to rotate the turbine, which improves the overall efficiency [68]. Another advantage of SOFC-APU systems in aircraft is the fuel reforming that can be performed directly within the SOFC [89–92]. Alternatively, SOFCs can utilize an external reformer that reuses the heat of the cells' electrochemical reaction [91, 93–97]. As SOFCs are more sulfur tolerant than PEMFCs, they can utilize various fuels for internal and external reforming processes.

Even though current SOFC-APU studies are predominantly theoretical, several companies market SOFC systems for transport applications. The SOFC technology maturity is at a TRL 9 for stationary applications; however, for transport applications, the TRL remains at 6 [98]. Table 4 provides some data on these companies and the SOFC characteristics that they developed.

**Table 3** SOFC-APU aircraft applications

| Power | Application |
|---|---|
| 5–240 kW | Regional mission airplanes and fully electric UAV [17, 18, 29, 59, 69–71] |
| Up to 500 kW | Short-range mission airplanes [21, 27, 72–74] |
| Higher than 400 kW | Long-range mission airplanes [40, 58, 75–83] |

Source: the authors

**Table 4** Example of SOFC developers and fuel cell application

| Company | Power | Fuel | Main application |
|---|---|---|---|
| Delphi (USA) | 9 kW/4 stacks have 403 $cm^2$ of active area | Natural gas, diesel, biodiesel, propane, gasoline, coal-derived fuel, and military logistics fuel | APU for trucks and heavy-duty commercial trucks |
| Fuelcell Energy (USA) | 1.4 mW | SOFC and solid oxide electrolyzer cell (SOEC) modes / coal gas, and natural gas | On-site power generation for large installations |
| Upstart power (USA) | 1000 W | Propane and natural gas | Residential, industrial, and remote power |
| Nexceris (USA) | On-demand | JP5 JP8, methane, ammonia, methanol, or hydrogen | Power generation, reversible for grid support |
| AVL (Austria) | 5–200 kW on-demand projects | Diesel, gasoline, ethanol, gas, and natural biofuels | Stationary power generator |
| Catator (Sweden) | 5 k | Natural gas, LPG,ethanol, methanol, diesel, jet fuel, and JP8 | Transport, balance of plant (bop) components for fuel cell systems |

Source: the authors

## 3.3 *Patents*

Fuel cells in aircraft are intensively developed among different industrial players, as they comprise a vital alternative to increase energy efficiency and reduce pollution. Aside from developments related to dedicated propulsion systems, a fuel cell APU is often considered a middle-term solution that can effectively increase overall efficiency and improve passengers' flight quality. Thus, there is a strong commercial interest in such developments, reflected in companies' patent activities. Patent data reveal valuable information on companies' commercial intention on specific technological development. This section focuses on PEMFC- and SOFC-APUs patent activity.

Figure 5 reports the patent filing from 1994 to 2019 from different companies that develop fuel cell APU technologies.[3] The commercial interest in fuel cell technologies for APUs started in the early 1990s when companies like Rockwell International, Lockheed, and Marconi Aerospace developed technologies to integrate these devices in aircraft. Later, in the early 2000s, Shimadzu and EUtech Scientific Engineering presented solutions to address fuel cell uses in aircraft, noticeably fuel cell water and humidity management and pressurization of the oxygen supplied to the fuel cell cathode.

In 2002, two major aircraft manufacturers, Airbus and Boeing started patenting their fuel cell APU technologies and intensified this technological trajectory,

[3] Data exclude patents related to electric propulsion aircraft and unmanned aerial vehicles.

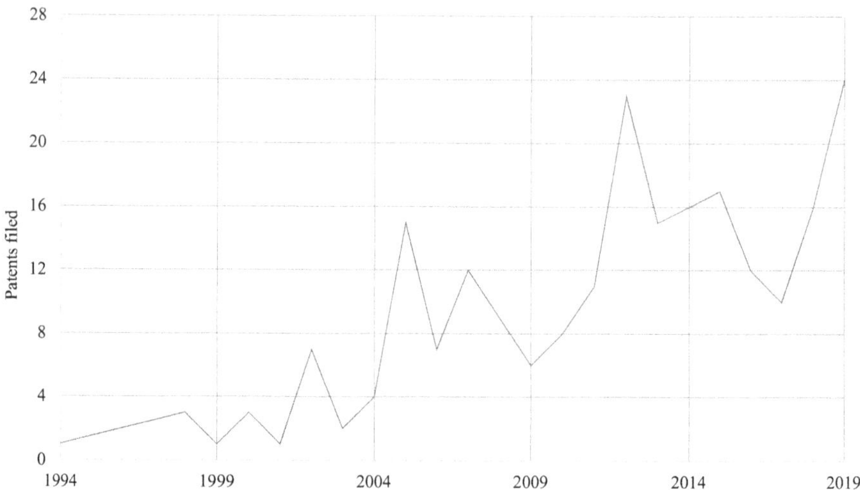


**Fig. 5** Fuel cell APU patent evolution from 1994 to 2019. (Source: the authors)

resulting in consistent patent filing after 2004. After 2010, patent activity increased even more and diversified due to component manufacturers entering the development and collaborating with the sector. This increased interaction led to a great diversity of alternatives to address fuel cell integration issues and related aircraft problems.

Table 5 shows the patent activity of fuel cell APU technologies and categorizes them according to the patent applicant. The data reveal that aircraft and aircraft component manufacturers are the leading developers of these technologies. Airbus alone holds 40% of all the patents identified, with Boeing following with 8%. These two significant manufacturers have different development strategies concerning the future of aviation. While Airbus includes alternatives to their portfolio to decarbonize and electrify aircraft, Boeing's strategy initially focuses on decarbonizing GPU and then moving on to decarbonizing aircraft by starting with APUs.

The patents aim at increasing the number of electric actuators and quality of passenger services offered by airlines (like catering, entertainment, hot water availability, and air conditioning), improving safety (inertization and emergency operation), and providing an additional heat source for water, air, and wing de-icing systems.

Component manufacturers often focus on patent specific parts such as the fuel tank, fuel cell operation methods, and BoP architecture, while major aircraft manufacturers focus on fuel cell integration. Patents related to fuel cell operations include fuel reforming, hydrogen production, oxygen supply to the cell in low pressure, exchanging fuel cell heat, managing fuel cell humidity and water, and converting direct current (DC) to alternating current (AC) for electric services for passengers like lighting or entertainment systems. To improve energy conversion efficiency and reduce weight, several patents disclose development on distributing fuel cell units within the aircraft, so the generated energy is locally utilized, for example, in

**Table 5** Patent categories according to company, number of patents, and patents per year

| Company | # | Patent per year |
|---|---|---|
| Airbus | 88 | 2002 (4), 2003 (1), 2004 (4), 2005 (12), 2006 (4), 2007 (9), 2008 (6), 2009 (4), 2010 (6), 2011 (4), 2012 (7), 2013 (8), 2014 (5), 2015 (3), 2016 (1), 2017 (2), 2018 (3), 2019 (4), 2020 (1) |
| Boeing | 17 | 2002 (3), 2005 (1), 2006 (2), 2008 (1), 2009 (1), 2012 (1), 2015 (2), 2016 (2), 2017 (1), 2018(2), 2019 (1) |
| Bell Helicopter Textron Inc | 10 | 2011 (1), 2017 (2), 2018 (1), 2019 (6) |
| Hamilton Sundstrand Corp | 7 | 2014 (1), 2015 (1), 2016 (1), 2018 (1), 2019 (3) |
| Zodiac Aerotechnics | 7 | 2012 (2), 2013 (2), 2014 (2), 2015 (1) |
| Ge Aviat Systems | 5 | 2012 (1), 2014 (1), 2015 (1), 2016 (1) |
| Honeywell Int Inc | 5 | 2007 (1), 2014 (2), 2015 (1), 2018 (1) |
| Liebherr Aerospace Gmbh | 5 | 2005 (1), 2008 (1), 2009 (1), 2011 (2) |
| Snecma | 5 | 2012 (2), 2014 (1), 2015 (2) |
| Bae Systems PLC | 4 | 2015 (1), 2017 (2), 2019 (1) |
| Diehl Aerospace Gmbh | 4 | 2007 (1), 2011 (1), 2012 (2) |
| Safran Power Units | 4 | 2014 (1), 2016 (2), 2017 (1) |
| Shimadzu Corp | 4 | 2000 (3), 2008 (1) |
| Parker Hannifin Corp | 3 | 2014 (1), 2015 (2) |
| Rolls Royce Corp | 3 | 2010 (1), 2011 (1), 2017 (1) |
| Commercial Aircraft Corp | 2 | 1999 (1), 2020 (1) |
| Dassault Aviat | 2 | 2012 (2) |
| Driessen Aerospace Group | 2 | 2012 (1), 2016 (1) |
| Eads Deutschland GMBH | 2 | 2011 (2) |
| GKN Aerospace Services Ltd | 2 | 2018 (1), 2019 (1) |
| Zeroavia Inc | 2 | 2019 (2) |
| Aircelle Sa | 1 | 2013 (1) |
| Avox Systems Inc | 1 | 2013 (1) |
| Eutech Scient Engineering | 1 | 2011 (1) |
| Honda Motor Co Ltd | 1 | 2018 (1) |
| Hylium Industry | 1 | 2019 (1) |
| Ihi Corp | 1 | 2012 (1) |
| Lockheed Corp | 1 | 1998 (1) |
| Lufthansa Technik Ag | 1 | 2013 (1) |

(continued)

**Table 5** (continued)

| Company | # | Patent per year |
|---|---|---|
| Mag Aerospace Ind Llc | 1 | 2012 (1) |
| Marconi Aerospace | 1 | 1998 (1) |
| Microturbo | 1 | 2013 (1) |
| Mitsubishi Heavy Industry | 1 | 2013 (1) |
| Rockwell International Corp | 1 | 1994 (1) |
| Shin Meiwa Ind Co Ltd | 1 | 2005 (1) |
| Siemens Ag | 1 | 2006 (1) |
| Solar Ship Inc | 1 | 2012 (1) |
| Suzhou Huatsing Power Sci.&Tech | 1 | 2014 (1) |
| Toyota | 1 | 2019 (1) |
| Turbomeka | 1 | 2006 (1) |
| Walmart Apollo | 1 | 2018 (1) |
| Xcellsis Gmbh | 1 | 1998 (1) |
| Zhejiang Hydrogen Tech | 1 | 2019 (1) |

Source: the authors

passenger seats, wings, or catering trollies. Moreover, greywater management addresses weight issues by reducing the water quantity necessary for fuel cell operation and aircraft passenger services.

More recent developments focus on integrating other renewable energy sources such as photovoltaic systems, combined with fuel cell APUs, and electrolyzes for on board hydrogen generation. Other inventions relate to, for example, removable hydrogen tanks, which reduce the dependency on the airport infrastructure, as the tank can be filled in a factory outside the airport.

## 4 Concluding Remarks

The increasing electricity demand and the search for higher efficiencies of energy conversion processes in aircraft are major pushes for using electrochemical devices such as chemical energy-to-electrical energy storage or converters on board. Fuel cell APUs are important alternatives proposed by research institutes and industrial players and their suppliers. Due to the increased output of patents and scientific papers, it is expected that such technologies will play a role in the aviation sector in the future, with enormous gains for society and the environment. Increased environmental constraints and passenger comfort are important factors motivating the development and implementation of fuel cell APUs and MEA.

**Acknowledgments** This study was financed in part by the Coordenação de Aperfeiçoamento de Pessoal de Nível Superior—Brasil (CAPES)—Finance Code 001, CNPQ, and FAPEMIG (Public Organization).

## References

1. IPCC (2019) Global Warming of 1.5°C. An IPCC Special Report on the impacts of global warming of 1.5°C above pre-industrial levels and related global greenhouse gas emission pathways, in the context of strengthening the global response to the threat of climate change,
2. Hydrogen Council (2017) Hydrogen scaling up: A sustainable pathway for the global energy transition
3. Bruce S, Temminghoff M, Hayward J, et al. (2020) Opportunities for hydrogen in aviation
4. Ganev E, Chiang C, Fizer L, Johnson E (2016) Electric Drives for Electric Green Taxiing Systems. SAE Int J Aerosp 9:62–73. https://doi.org/10.4271/2016-01-2013
5. Kadyk T, Winnefeld C, Hanke-Rauschenbach R, Krewer U (2018) Analysis and Design of Fuel Cell Systems for Aviation. Energies 11:1–15. https://doi.org/10.3390/en11020375
6. Hussain AM, Wachsman ED (2019) Liquids-to-Power Using Low-Temperature Solid Oxide Fuel Cells. Energy Technol 7:20–32. https://doi.org/10.1002/ente.201800408
7. Sgroi MF, Zedde F, Barbera O, et al. (2016) Cost analysis of direct methanol fuel cell stacks for mass production. Energies 9:. https://doi.org/10.3390/en9121008
8. ACRP (2012) Handbook for Evaluating Emissions and Costs of APUs and Alternative Systems. Transportation Research Board, Washington, D.C.
9. Jackson P, Munson K (2004) Jane's All the World's Aircraft (2004-2005). Jane's Information Group
10. SAE (2021) Aerospace Auxiliary Power Sources. AIR744
11. Sarlioglu B, Morris CT (2015) More Electric Aircraft: Review, Challenges, and Opportunities for Commercial Transport Aircraft. IEEE Trans Transp Electrif 1:54–64. https://doi.org/10.1109/TTE.2015.2426499
12. Freeh JE (2014) Hydrogen Fuel Cells for Auxiliary Power Units. Encycl Aerosp Eng 1–7. https://doi.org/10.1002/9780470686652.eae1028
13. Cruz Champion H, Kabelac S (2017) Multifunctional fuel cell system for civil aircraft: Study of the cathode exhaust gas dehumidification. Int J Hydrogen Energy 42:29518–29531. https://doi.org/10.1016/j.ijhydene.2017.09.175
14. Masiol M, Harrison RM (2014) Aircraft engine exhaust emissions and other airport-related contributions to ambient air pollution: A review. Atmos Environ 95:409–455. https://doi.org/10.1016/j.atmosenv.2014.05.070
15. Moreno RF, Economou JT, Bray D, Knowles K (2013) Modelling and simulation of a fuel cell powered electric drivetrain for wide body passenger aircraft. Proc Inst Mech Eng Part G J Aerosp Eng 227:608–617. https://doi.org/10.1177/0954410012473389
16. Thalin P, Rajamani R, Mare J-C, Taubert S (2018) Fundamentals of Electric Aircraft
17. Choudhary T, Sahu MK, Sanjay R, et al. (2018) Thermodynamic Modeling of Blade Cooled Turboprop Engine Integrated to Solid Oxide Fuel Cell: A Concept. SAE Tech Pap 2018-April:1–10. https://doi.org/10.4271/2018-01-1308
18. Choudhary T, Sahu M, Krishna S (2017) Thermodynamic Analysis of Solid Oxide Fuel Cell Gas Turbine Hybrid System for Aircraft Power Generation. SAE Tech Pap 2017-Septe: https://doi.org/10.4271/2017-01-2062
19. Peters R, Samsun RC (2013) Evaluation of multifunctional fuel cell systems in aviation using a multistep process analysis methodology. Appl Energy 111:46–63. https://doi.org/10.1016/j.apenergy.2013.04.058

20. Thirkell A, Chen R, Harrington I (2017) A Fuel Cell System Sizing Tool Based on Current Production Aircraft. SAE Tech Pap Part F1298: https://doi.org/10.4271/2017-01-2135
21. Braun RJ, Gummalla M, Yamanis J (2009) System architectures for solid oxide fuel cell-based auxiliary power units in future commercial aircraft applications. J Fuel Cell Sci Technol 6:0310151–03101510. https://doi.org/10.1115/1.3008037
22. Boeing Commercial Airplanes (2018) 787 Airplane Characteristics for Airport Planning - D6-58333 - Revision M
23. Frew S, Fraga J (2007) 787 No-Bleed Systems. Boeing Aero Mag. Q4-07 32
24. Behling NH (2013) Introduction. Fuel Cells 1–6. https://doi.org/10.1016/b978-0-444-56325-5.00001-6
25. Burke K (2012) Current perspective on hydrogen and fuel cells. Elsevier Ltd.
26. Lardier C (2018) The soviet manned lunar program N1-L3. Acta Astronaut 142:184–192. https://doi.org/10.1016/j.actaastro.2017.10.007
27. Gummalla M, Pandy A, Braun R, et al. (2006) Fuel Cell Airframe Integration Study for Short-Range Aircraft NASA STI Program. Volume 1:
28. Keim M, Kallo J, Friedrich KA, et al. (2013) Multifunctional fuel cell system in an aircraft environment: An investigation focusing on fuel tank inerting and water generation. Aerosp Sci Technol 29:330–338. https://doi.org/10.1016/j.ast.2013.04.004
29. Santarelli M, Cabrera M, Calì M (2009) Analysis of solid oxide fuel cell systems for more-electric aircraft. J Aircr 46:269–283. https://doi.org/10.2514/1.38408
30. Turpin C, Morin B, Bru E, et al. (2017) Power for Aircraft Emergencies: A Hybrid Proton-Exchange Membrane H2\/O2 Fuel Cell and Ultracapacitor System. IEEE Electrif Mag 5:72–85. https://doi.org/10.1109/MELE.2017.2758879
31. SAE (2017) Installation of Fuel Cell Systems in Large Civil Aircraft AS6858
32. Pratt JW, Brouwer J, Samuelsen GS (2007) Performance of proton exchange membrane fuel cell at high-altitude conditions. J Propuls Power 23:437–444. https://doi.org/10.2514/1.20535
33. Whyatt GA, Chick LA (2012) Electrical Generation for More-Electric Aircraft Using Solid Oxide Fuel Cells. US Dep Energy 110
34. Guida D, Minutillo M (2017) Design methodology for a PEM fuel cell power system in a more electrical aircraft. Appl Energy 192:446–456. https://doi.org/10.1016/j.apenergy.2016.10.090
35. Pratt JW, Klebanoff LE, Munoz-Ramos K, et al. (2011) Proton exchange membrane fuel cells for electrical power generation on-board commercial airplanes
36. Bégot S, Harel F, Candusso D, et al (2010) Fuel cell climatic tests designed for new configured aircraft application. Energy Convers Manag 51:1522–1535. https://doi.org/10.1016/j.enconman.2010.02.011
37. Campanari S, Manzolini G, Beretti A, Wollrab U (2008) Performance assessment of turbocharged pem fuel cell systems for civil aircraft on-board power production. J Eng Gas Turbines Power 130:. https://doi.org/10.1115/1.2772636
38. Romeo G, Correa G, Borello F, et al. (2012) Air Cooling of a Two-Seater Fuel Cell–Powered Aircraft: Dynamic Modeling and Comparison with Experimental Data. J Aerosp Eng 25:356–368. https://doi.org/10.1061/(asce)as.1943-5525.0000138
39. Daggett DL, Eelman S, Kristiansson G (2003) Fuel cell APU for commercial aircraft. AIAA\ICAS Int Air Sp Symp Expo Next 100 Years 1–9
40. Rajashekara K, Grieve J, Daggett D (2008) Hybrid fuel cell power in aircraft. IEEE Ind Appl Mag 14:54–60. https://doi.org/10.1109/MIAS.2008.923606
41. Santarelli M, Cabrera M, Calí M (2010) Solid oxide fuel based auxiliary power unit for regional jets: Design and mission simulation with different cell geometries. J Fuel Cell Sci Technol 7:0210061–02100611. https://doi.org/10.1115/1.3176282
42. SAE (2021) Aerospace Fuel System Specifications and Standards. AIR1408
43. SAE (2019) Considerations for Hydrogen Fuel Cells in Airborne Applications. AIR7765
44. SAE (2020) EUROCAE/SAE WG80/AE-7AFC Hydrogen Fuel Cells Aircraft Fuel Cell Safety Guidelines. AIR6464

45. Chick LA (2013) Cost Study for Manufacturing of Solid Oxide Fuel Cell Power Systems MR Weimar LA Chick DW Gotthold GA Whyatt. 50
46. Westenberger A (2016) Hydrogen and fuel cell: Technologies and market perspectives. In: Töpler J, Lehmann J (eds) Hydrogen and Fuel Cell. pp 107–125
47. Kallo J, Renouard-vallet G, Saballus M, et al. (2010) Fuel Cell System Development and Testing for Aircraft Appli- cations Fuel Cell System Development and Testing for Aircraft Applications. 18thWorld Hydrog Energy Conf 2010 - WHEC 2010 78:435–444
48. Dyantyi N, Parsons A, Bujlo P, Pasupathi S (2019) Behavioural study of PEMFC during start-up/shutdown cycling for aeronautic applications. Mater Renew Sustain Energy 8:1–8. https://doi.org/10.1007/s40243-019-0141-4
49. Hordé T, Achard P, Metkemeijer R (2012) PEMFC application for aviation: Experimental and numerical study of sensitivity to altitude. Int J Hydrogen Energy 37:10818–10829. https://doi.org/10.1016/j.ijhydene.2012.04.085
50. Novillo E, Pardo M, García-Luis A (2010) Novel approaches for the integration of high tem perature PEM fuel cells into aircrafts. ASME 2010 8th Int Conf Fuel Cell Sci Eng Technol FUELCELL 2010 2:479–487. https://doi.org/10.1115/FuelCell2010-33090
51. Werner C, Preiß G, Gores F, et al. (2016) A comparison of low-pressure and supercharged operation of polymer electrolyte membrane fuel cell systems for aircraft applications. Prog Aerosp Sci 85:51–64. https://doi.org/10.1016/j.paerosci.2016.07.005
52. Brooks N, Baldwin T, Brinson T, et al. (2004) Analysis of fuel cell based power systems using EMTDC electrical power simulator. Proc Annu Southeast Symp Syst Theory 36:270–274. https://doi.org/10.1109/ssst.2004.1295662
53. Belmonte N, Staulo S, Fiorot S, et al. (2018) Fuel cell powered octocopter for inspection of mobile cranes: Design, cost analysis and environmental impacts. Appl Energy 215:556–565. https://doi.org/10.1016/j.apenergy.2018.02.072
54. Ogungbemi E, Wilberforce T, Ijaodola O, et al. (2020) Selection of proton exchange membrane fuel cell for transportation. Int J Hydrogen Energy. https://doi.org/10.1016/j.ijhydene.2020.06.147
55. Samsun RC, Pasel J, Janßen H, et al. (2014) Design and test of a 5kWe high-temperature polymer electrolyte fuel cell system operated with diesel and kerosene. Appl Energy 114:238–249. https://doi.org/10.1016/j.apenergy.2013.09.054
56. Renouard-Vallet G, Saballus M, Schumann P, et al. (2012) Fuel cells for civil aircraft application: On-board production of power, water and inert gas. Chem Eng Res Des 90:3–10. https://doi.org/10.1016/j.cherd.2011.07.016
57. Youngblood JW, Talay TA, Pegg RJ (1984) Design of long-endurance unmanned airplanes incorporating solar and fuel cell propulsion. In: AIAA//SAE/ASEE 20th Joint Propulsion Conference. Cincinnati, Ohio
58. Eelman S, del Pozo y de Poza I, Krieg T (2004) Fuel Cell APU's in Commercial Aircraft – an Assessment of SOFC and PEMFC Concepts. 24th Int Congr Aeronaut Sci (IC 1–10
59. Sehra AK, Whitlow W (2004) Propulsion and power for 21st century aviation. Prog Aerosp Sci 40:199–235. https://doi.org/10.1016/j.paerosci.2004.06.003
60. Buonomano A, Calise F, d'Accadia MD, et al. (2015) Hybrid solid oxide fuel cells-gas turbine systems for combined heat and power: A review. Appl Energy 156:32–85. https://doi.org/10.1016/j.apenergy.2015.06.027
61. Gengo T, Kobayashi Y, Hisatome N, et al (2007) Progressing Steadily, Development of High-Efficiency SOFC Combined Cycle System. Mitsubishi Heavy Ind Ltd Tech Rev 44:1–5
62. Lim TH, Song RH, Shin DR, et al. (2008) Operating characteristics of a 5 kW class anode-supported planar SOFC stack for a fuel cell/gas turbine hybrid system. Int J Hydrogen Energy 33:1076–1083. https://doi.org/10.1016/j.ijhydene.2007.11.017
63. Seidler S, Henke M, Kallo J, et al. (2011) Pressurized solid oxide fuel cells: Experimental studies and modeling. J Power Sources 196:7195–7202. https://doi.org/10.1016/j.jpowsour.2010.09.100

64. Rechberger J, Kaupert A, Hagerskans J, Blum L (2016) Demonstration of the First European SOFC APU on a Heavy Duty Truck. Transp Res Procedia 14:3676–3685. https://doi.org/10.1016/j.trpro.2016.05.442
65. Reeve N (2016) Nissan unveils world's first Solid-Oxide Fuel Cell vehicle. In: Nissan Off. Glob. Newsroom. https://global.nissannews.com/en/releases/nissan-unveils-worlds-first-solid-oxide-fuel-cell-vehicle
66. Sunfire (2015) Sunfire supplied Thyssenkrupp marine systems with 50 kW SOFC
67. Tse LKC, Wilkins S, McGlashan N, et al. (2011) Solid oxide fuel cell/gas turbine trigeneration system for marine applications. J Power Sources 196:3149–3162. https://doi.org/10.1016/j.jpowsour.2010.11.099
68. Fernandes MD, de ST, Bistritzki VN, et al. (2018) SOFC-APU systems for aircraft: A review. Int J Hydrogen Energy 43:16311–16333. https://doi.org/10.1016/j.ijhydene.2018.07.004
69. Aguiar P, Brett DJL, Brandon NP (2008) Solid oxide fuel cell/gas turbine hybrid system analysis for high-altitude long-endurance unmanned aerial vehicles. Int J Hydrogen Energy 33:7214–7223. https://doi.org/10.1016/j.ijhydene.2008.09.012
70. Fernandes A, Woudstra T, Aravind P V. (2015) System simulation and exergy analysis on the use of biomass-derived liquid-hydrogen for SOFC/GT powered aircraft. Int J Hydrogen Energy 40:4683–4697. https://doi.org/10.1016/j.ijhydene.2015.01.136
71. Waters DF, Cadou CP (2015) Engine-integrated solid oxide fuel cells for efficient electrical power generation on aircraft. J Power Sources 284:588–605. https://doi.org/10.1016/j.jpowsour.2015.02.108
72. Yanovskiy LS (2012) Fuel cells on alternative non-oil fuels for gas turbine engines of an advanced civil aircrafts. 28th Congr Int Counc Aeronaut Sci 2012, ICAS 2012 4:2921–2926
73. Fateh S (2016) Bi-directional Solid Oxide Cells used as SOFC for Aircraft APU system and as SOEC to produce fuel at the airport. TU Delft
74. Tunca F, Kaya N (2017) Thermodynamic Analysis of Gas Turbine – Solid Oxide Fuel Cell (GT-SOFC) Aircraft Auxiliary Power Unit (APU). Int J Adv Mech Automob Eng 4:10–14. https://doi.org/10.15242/ijamae.iae1216202
75. Nehter P, Winkler WG (2015) System analysis of Fuel Cell APUs for Aircraft applications. In: h2expo Hamburg
76. Dollmayer J, Bundschuh N, Carl UB (2006) Fuel mass penalty due to generators and fuel cells as energy source of the all-electric aircraft. Aerosp Sci Technol 10:686–694. https://doi.org/10.1016/j.ast.2006.08.001
77. Steffen CJ, Freeh JE, Larosiliere LM (2005) Solid oxide fuel cell/gas turbine hybrid cycle technology for auxiliary aerospace power. Proc ASME Turbo Expo 5:253–260. https://doi.org/10.1115/GT2005-68619
78. Srinivasan H, Yamanis J, Welch R, et al. (2006) Solid Oxide Fuel Cell APU Feasibility Study for a Long Range Commercial Aircraft Using UTC ITAPS Approach Volume I : Aircraft Propulsion and Subsystems Integration Evaluation. Nasa Cr--2006-211458 1:
79. Tornabene R, Wang XY, Steffen CJ, Freeh JE (2005) Development of parametric mass and volume models for an aerospace SOFC/gas turbine hybrid system. Proc ASME Turbo Expo 5:135–144. https://doi.org/10.1115/GT2005-68334
80. Freeh JE, Steffen CJ, Larosiliere LM (2005) OFF-DESIGN PERFORMANCE ANALYSIS OF A SOLID-OXIDE FUEL CELL/GAS TURBINE HYBRID FOR AUXILIARY AEROSPACE POWER. In: Third International Conference on Fuel Cell Science, Engineering and Technology. pp 1–8
81. Chinda P, Brault P (2012) The hybrid solid oxide fuel cell (SOFC) and gas turbine (GT) systems steady state modeling. Int J Hydrogen Energy 37:9237–9248. https://doi.org/10.1016/j.ijhydene.2012.03.005
82. Evrin RA, Dincer I (2020) Development and evaluation of an integrated solid oxide fuel cell system for medium airplanes. Int J Energy Res 44:9674–9685. https://doi.org/10.1002/er.5525

83. Singh AS, Choudhary T, Sanjay S (2019) Thermal Analysis of Aircraft Auxiliary Power Unit: Potential of Super-Critical CO2 Brayton Cycle. SAE Tech Pap 1–12. https://doi.org/10.4271/2019-01-1391
84. Wu S, Li Y (2014) Fuel cell applications on more electrical aircraft. 17th Int Conf Electr Mach Syst ICEMS 198–201. https://doi.org/10.1109/ICEMS.2014.7013481
85. Barelli L, Bidini G, Ottaviano A (2013) Part load operation of a SOFC/GT hybrid system: Dynamic analysis. Appl Energy 110:173–189. https://doi.org/10.1016/j.apenergy.2013.04.011
86. Barelli L, Bidini G, Ottaviano A (2012) Part load operation of SOFC/GT hybrid systems: Stationary analysis. Int J Hydrogen Energy 37:16140–16150. https://doi.org/10.1016/j.ijhydene.2012.08.015
87. Costamagna P, Magistri L, Massardo AF (2001) Design and part-load performance of a hybrid system based on a solid oxide fuel cell reactor and a micro gas turbine. J Power Sources 96:352–368. https://doi.org/10.1016/S0378-7753(00)00668-6
88. Matencio T, Domingues RZ, Andrade ST de P, et al. (2017) Modeling of an airship auxiliary power unit based on solid oxide fuel cells. In: Hypothesis XII. Sircacusa, Italy, p 120
89. Laosiripojana N, Assabumrungrat S (2007) Catalytic steam reforming of methane, methanol, and ethanol over Ni/YSZ: The possible use of these fuels in internal reforming SOFC. J Power Sources 163:943–951. https://doi.org/10.1016/j.jpowsour.2006.10.006
90. Liso V, Olesen AC, Nielsen MP, Kær SK (2011) Performance comparison between partial oxidation and methane steam reforming processes for solid oxide fuel cell (SOFC) micro combined heat and power (CHP) system. Energy 36:4216–4226. https://doi.org/10.1016/j.energy.2011.04.022
91. Santin M, Traverso A, Magistri L, Massardo A (2010) Thermoeconomic analysis of SOFC-GT hybrid systems fed by liquid fuels. Energy 35:1077–1083. https://doi.org/10.1016/j.energy.2009.06.012
92. Shiratori Y, Ijichi T, Oshima T, Sasaki K (2010) Internal reforming SOFC running on biogas. Int J Hydrogen Energy 35:7905–7912. https://doi.org/10.1016/j.ijhydene.2010.05.064
93. Aicher T, Lenz B, Gschnell F, et al (2006) Fuel processors for fuel cell APU applications. J Power Sources 154:503–508. https://doi.org/10.1016/j.jpowsour.2005.10.026
94. Cocco D, Tola V (2009) Externally reformed solid oxide fuel cell-micro-gas turbine (SOFC-MGT) hybrid systems fueled by methanol and di-methyl-ether (DME). Energy 34:2124–2130. https://doi.org/10.1016/j.energy.2008.09.013
95. Shi L, Bayless DJ (2008) Analysis of jet fuel reforming for solid oxide fuel cell applications in auxiliary power units. Int J Hydrogen Energy 33:1067–1075. https://doi.org/10.1016/j.ijhydene.2007.11.012
96. Triphob N, Wongsakulphasatch S, Kiatkittipong W, et al. (2012) Integrated methane decomposition and solid oxide fuel cell for efficient electrical power generation and carbon capture. Chem Eng Res Des 90:2223–2234. https://doi.org/10.1016/j.cherd.2012.05.014
97. Valadez Huerta G, Álvarez Jordán J, Dragon M, et al. (2018) Exergy analysis of the diesel pre-reforming solid oxide fuel cell system with anode off-gas recycling in the SchIBZ project. Part I: Modeling and validation. Int J Hydrogen Energy 43:16684–16693. https://doi.org/10.1016/j.ijhydene.2018.04.216
98. U.S. Department of Energy (2019) Report on the Status of the Solid Oxide Fuel Cell Program

# Solid Oxide Fuel Cell Systems and Their Potential Applications in the Aviation Industry and Beyond

**Vikrant Venkataraman**

## 1 Introduction

The world is definitely making a transition to alternate fuels and slowly leaving fossil fuels behind. It is envisaged that in the coming decades, an alternate or a mix of different fuels will take over the role of providing energy and power to various sectors. Among the different fuels, hydrogen is touted to be one of the options, and there is a lot of emphasis on building a hydrogen economy in certain countries like Japan, Germany and China. The former two have kick started their program much earlier than China, and it is only in the last 5 years that China has given a strong push for hydrogen and fuel cells. There is an interest in the hydrogen economy and activities in other countries as well but only in certain dedicated pockets, and there is no strong national agenda to adopt hydrogen or fuel cell technology as the economies of scale are still not attractive. However, recently, many European countries have come out a dedicated hydrogen strategy at a national level which has in turn made the EU (European Union) come out with a hydrogen strategy. These hydrogen strategies lay down the vision of these countries, what they would like to achieve and the amount of money they are willing to put into.

The hydrogen economy, if completely developed, could be utopia for mankind. A scenario where hydrogen is produced from renewable sources and is used as the fuel in all major sectors, thereby leading to almost zero or very minimal emissions. However, in order to get to an economy based on hydrogen from the current fossil fuel-based economy, a transition step is required. Sudden step changes cannot be done as industries are sluggish to respond to step changes, and new technologies (based on different fuels) should be able to provide the same functions/functionality

V. Venkataraman (✉)
Fuel Cells – Instrumentation and Test Systems, AVL List GmbH, Graz, Austria
e-mail: vikrant.venkataraman@avl.com

 
C. O. Colpan, A. Kovač (eds.), *Fuel Cell and Hydrogen Technologies in Aviation*, Sustainable Aviation, https://doi.org/10.1007/978-3-030-99018-3_8

and user experience as the current technologies with minimal or no disruption. Of late, there has also been a lot of emphasis on generation of e-fuels (electro-fuels) which are made from $CO_2$ capture along with hydrogen produced from renewable energy. The fuels made in this manner can use the existing fuel infrastructure network for storage and distribution.

Hydrogen could definitely revolutionize the transport sector, all forms of transport (road, rail and water) could be running on hydrogen or its derivatives. Hydrogen powering the commercial aviation sector might not be realized in the near future although there are predictions for an all-electric aircraft (which can perform with similar functionalities as the current aircrafts) by the year 2040. Hydrogen is just one of the fuel options in the long-term vehicle mix. The world could also easily transition to biofuels, e-fuels, methanol or even ammonia.

Polymer electrolyte fuel cells (PEFCs) have already started taking the role as the prime mover in passenger cars, examples include the Toyota MIRAI, Honda Clarity and the Hyundai ix35. Many models from other OEMs are also predicted to enter the market in the next decade. The PEFC relies on a hydrogen-based economy though. However, other major forms of transportation (heavy duty vehicles, ships and airplanes) have not yet moved out from conventional fossil fuels, and this presents a great challenge if one wants to decarbonize the transport sector. What is certain in the current decade is that powertrain electrification is on the rise but whether it will solely depend on fuel cell or battery technology or a mix of both, only time will tell.

Fossil fuels are still going to linger around for at least three to four decades and will account for 60% of the total transport fuel demand [1] but how can we best make use of them without increasing emissions and at the same time increase energy efficiency of the systems that use them? In the road leading to hydrogen, using other fuel cell technologies that can still operate on fossil fuels but with better efficiencies and lower emissions compared to internal combustion engines (ICE) will lay down a path when a full-fledged hydrogen economy develops. Scenarios from the recent IEA (International Energy Agency) hydrogen and fuel cells roadmap suggest a 30% share for hydrogen vehicles by 2050 [2]. Now this is still not sufficient as the remaining 70% of the vehicles would be still running on conventional fuels. Solid oxide fuel cell (SOFCs) which fall into the class of high temperature fuel cells could potentially fill this gap and help make the transition for the transportation sector to completely move away from ICE technology. Figure 1 helps depict the point just stated.

Hydrogen as a fuel might be suitable for passenger cars however for heavy duty vehicles such as trucks, shipping vessels and airplanes, the tank size required might be too big and heavy and that's where SOFCs come to the rescue as they can operate on different kinds of fuels and are thus termed flex-fuel devices. The incorporation of SOFCs in the transport sector along with PEFCs will give sufficient confidence to the fuel cell industry and the hydrogen industry alike to push forward with this technology in general.

This chapter is intended to provide the reader a flavour of Solid Oxide Fuel Cells and systems built around it and also potential application of such systems in the aerospace industry as the focus of the book is on employing fuel cell technologies for aviation. The chapter is organized as follows:

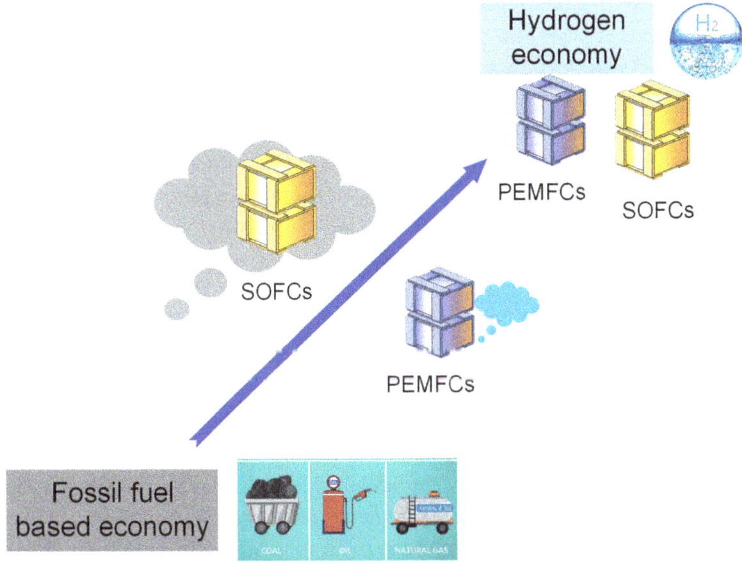


**Fig. 1** Use of SOFCs in transitioning to a $H_2$-based economy

*Section 2* provides a brief description of what an SOFC is and how it functions, the different stack and cell designs available.

*Section 3* focuses on the different fuels which can potentially be used in SOFCs for the most promising applications envisaged in the next decade.

*Section 4* will touch on two SOFC system architectures, one running on hydrogen and the other running on methane. This will detail out the major system components involved and also explain the possibilities of having different architectures depending on the end application.

*Section 5* will talk about the needs of the aviation industry in terms of power and heat requirements and where in the aviation sector, airplanes and airports included, can SOFCs be potentially employed.

*Section 6* will touch upon the concluding remarks and the perspectives as to what one can expect in the future in this field with SOFC technology.

## 2 What Is an SOFC and How Does It Work?

A solid oxide fuel cell is a device which works in the temperature range of 600–1000 °C. It falls under the category of *high temperature fuel cells*. Recently, the SOFCs have been further sub-classified into 'high temperature', meaning working greater than 700 °C and 'intermediate temperature', meaning working between 600 and 700 °C. As the fuel is electrochemically combusted, there is no nitrous oxide ($NO_x$) or fine particulate matter in the exhaust emissions from an SOFC-based system. Also, as sulphur is a poison for the catalyst of the SOFC, sulphur oxide ($SO_x$)

emissions are almost insignificant. This is because the fuel is pre-reformed and cleaned to an extent prior to feeding it to the SOFC. The water produced in an SOFC is on the fuel electrode side and is entirely gaseous in nature. This helps in steam reforming of fuels on the electrode itself which is an advantage because reforming is generally an endothermic process, and fuel cell operation is an exothermic one. Thus technically, internal reforming can also aid in thermal management of the stack, thereby satisfying the thermal needs of both processes.

The basic reaction (with hydrogen as fuel) happening on the fuel electrode and air electrode is presented in Eqs. (1) and (2), respectively.

$$\text{Fuel electrode}\quad H_2 + O^{2-} \rightarrow H_2O + 2e^- \tag{1}$$

$$\text{Air electrode}\quad O_2 + 2e^- \rightarrow O^{2-} \tag{2}$$

Note: *The terms anode and cathode are not used here because the solid oxide cell can also be used in electrolysis mode, and in that case, the terms get reversed. Hence, to avoid confusion, the terms fuel electrode and air electrode are used.*

So basically, oxygen gets reduced to oxide ions at the air electrode, these oxide ions then permeate through the membrane and reach the other side where they combine with the fuel and electrochemical combustion occurs, resulting in the release of electrons and generation of water vapour. These electrons then flow through the external circuit and reach the air electrode, repeating the process once again.

With fuels other than hydrogen, the reactions at the fuel electrode are more complex and not as straightforward as the one described in Eq. (1). As an example, the set of reactions involved when methane (or natural gas) is used as the fuel is given below.

$$CH_4 + H_2O \rightarrow CO + 3H_2 \tag{3}$$

$$CO + H_2O \rightarrow CO_2 + H_2 \tag{4}$$

$$CH_4 + 2H_2O \rightarrow CO_2 + 4H_2 \tag{5}$$

Equation (3) is the basic steam methane reforming reaction where the $CH_4$ species is converted to CO (carbon monoxide) and $H_2$. The CO then reacts via the water gas shift reaction mentioned in Eq. (4), to generate additional hydrogen and $CO_2$. The overall reaction is summarized in Eq. (5). Thus, one mole of methane generates four moles of hydrogen when reformed.

## 2.1 Different Cell Designs

There are two types of SOFCs—tubular and planar, and they are further classified as:

- Tubular (diameter > 15 mm)

  - m-tubular (diameter < 5 mm)
- Flat tubular.[1]
- Planar—square or circular

A schematic representation of these types of fuel cells is presented in Fig. 2.

Planar designs in general offer higher power density when compared to tubular designs. However, the tubular designs have higher volumetric power density and have a low thermal mass which in turn translates to higher resistance to thermal shock, rapid turn on/off capabilities and relatively longer life due to operation at lower current densities. Tubular SOFCs have relatively lesser sealing problems compared to planar SOFCs because the fuel and air go inside and outside the tube (or vice versa depending on which electrode falls inside or outside). Planar SOFCs on the other hand use glass seals and other glassy type seals which may degrade over time at high temperature. Thus, the sealing between the fuel and air electrodes is more critical in a planar design when compared to a tubular design and therefore tubular geometries have an advantage over planar ones with regard to seals.

Tubular SOFCs have been built for multi-kilowatt power levels by Siemens Westinghouse and currently planar SOFCs have also been built for several kWs by companies like Bloom Energy and Fuel Cell energy. As of today, planar SOFCs seem to be the preferred design choice with most SOFC stack manufacturers.

## 2.2 *Different Cell Architectures*

Besides different stack designs, the single cell of the SOFC can have different designs too viz.:

- Electrolyte supported or
- Fuel electrode supported (anode supported) or
- Air electrode supported (cathode supported)

Making the electrolyte as thin as possible reduces the ohmic losses, thereby increasing the ionic conductivity and allows the cell/stack to operate at a slightly lower temperature. A thicker electrolyte would warrant the cell/stack to operate at higher temperatures. The reason for this is the direct correlation between ohmic losses and thickness of electrolyte and also the temperature dependence of the conductivity of the electrolyte. This is the reason why many SOFC cell/stack manufacturers opt for fuel electrode or air electrode supported designs, the former being more common. The different kind of cell architectures is shown in Fig. 3.

[1] LG manufacturers these kind of cells.

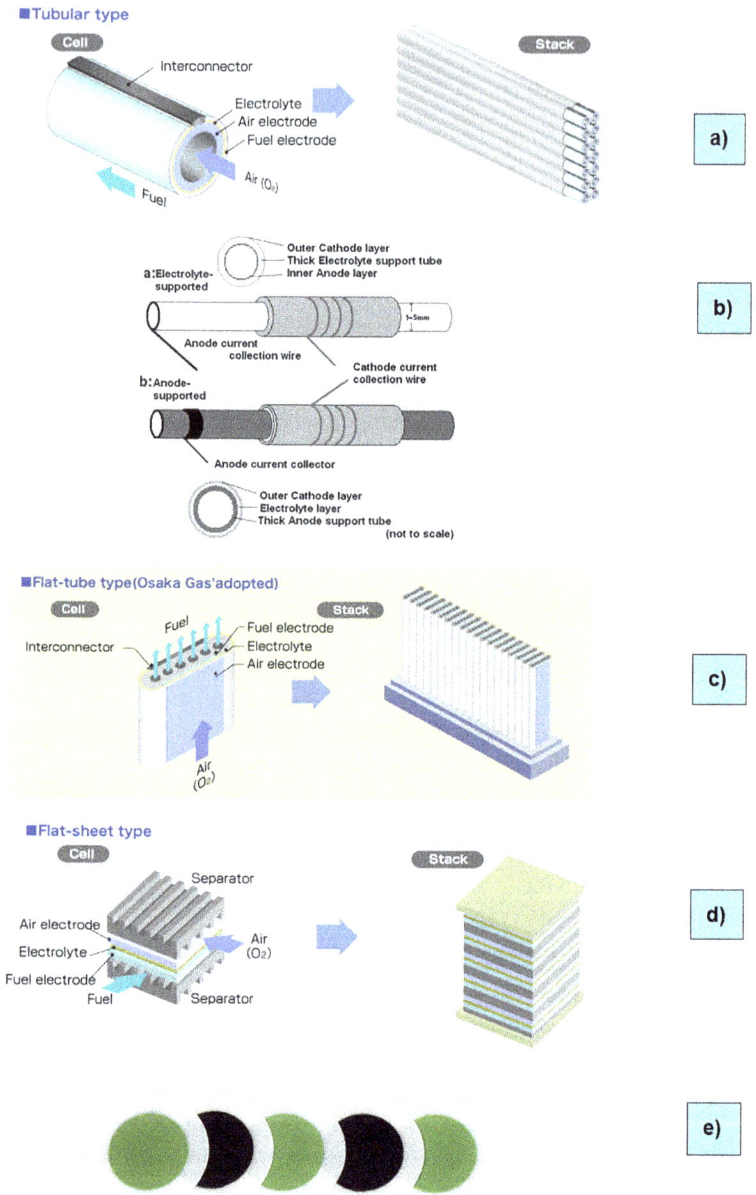


**Fig. 2** Different designs of solid oxide fuel cells: (**a**) tubular, (**b**) micro-tubular, (**c**) flat tubular, (**d**) planar square, (**e**) planar circular. (Source: (**a**) Tubular, (**c**) flat tubular and (**d**) planar: https://www.osakagas.co.jp/en/rd/fuelcell/sofc/sofc/system.html); Source: Micro-tubular: Journal of Power Sources, vol 196, 1677–1686, 2011; Source: planar circular: https://fuelcellmaterials.com/nextcell-hp-bringing-advanced-manufacturing-to-sofc-components-for-improved-performance)

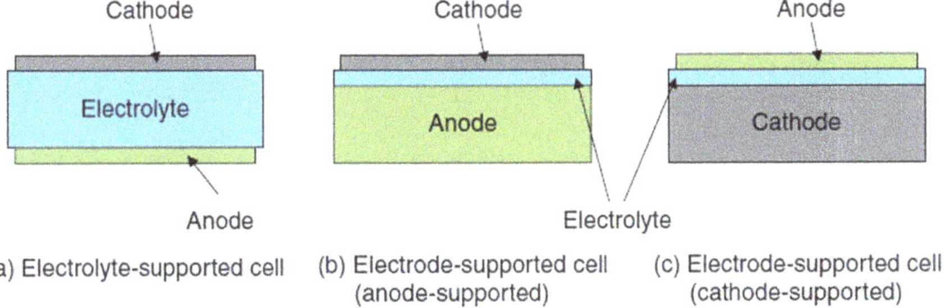


**Fig. 3** Different single-cell architectures for SOFCs [3]

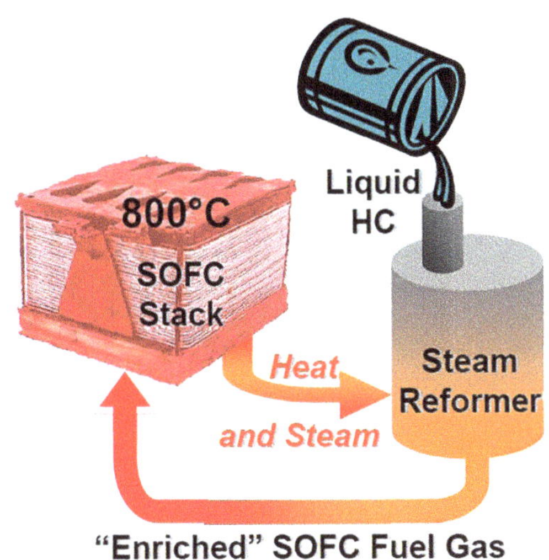


**Fig. 4** Schematic showing how SOFC can be used with current fuel mix [4]. (*Liquid HC* liquid hydrocarbons)

## 2.3 *Advantages of SOFC over PEMFCs*

So why should one use SOFCs over PEFCs? Some of the key points that give SOFC an edge over the PEFC are:

1. Fuel flexibility—technically any of these fuels—diesel, methane, methanol, syngas, ammonia, bio-ethanol, bio-methanol can be used (with appropriate reforming and gas cleaning systems in place). A basic schematic of this is shown in Fig. 4.
2. Is able to fit in very well with the current fuel infrastructure. This can play a very crucial role in adoption of the technology by industry.
3. Higher tolerance to impurities. The tolerance to sulphur is 10 ppm with SOFCs, whereas it is less than 1 ppm for PEMFCs. Carbon deposition on catalytic sites will immediately kill the PEFC, but the SOFC can work for a few hours. It is also

predicted that carbon deposition can be eliminated if steam (or oxygen containing species) is passed through the fuel electrode.

4. Very high electrical efficiencies.[2]
5. CHP (combined heat and power) capability, reaching an efficiency of >80%.
6. No water management, thus avoiding complex thermal management and coolant loops.
   (a) In hot climates, the PEFC system may not be able to reject sufficient heat to the ambient, thereby resulting in high stack temperatures which might lead to system failure.
7. Does not use expensive catalysts such as platinum or combinations of platinum.

## 3 Potential Fuels for SOFCs

As mentioned earlier, SOFCs are able to operate on a range of different fuels. The most promising applications for SOFCs in the next decade and the potential fuels that will be used for these applications are shown in Fig. 5.

Natural gas (or methane) is the common fuel among all applications (using SOFCs) and thus will be the fuel of choice until a complete hydrogen economy kicks in. So what is the natural gas grid of today might be converted to a hydrogen

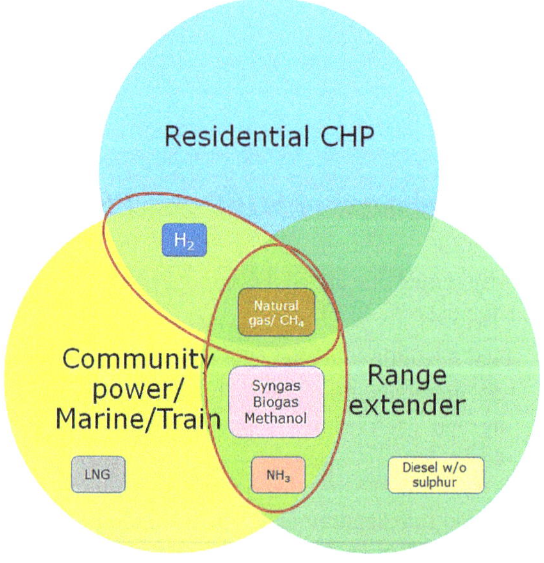


**Fig. 5** Venn diagram showing most promising applications till 2030 and fuels for those applications

[2] SOFCMAN, China have reported electrical efficiencies of 72.5% (LHV) for their stacks. The average electrical efficiency for an SOFC stack running on hydrogen is roughly 60%. Elcogen stacks have achieved a world record efficiency of 74%.

gas grid of tomorrow, and if SOFC systems are deployed for stationary applications, then they will be connected to the natural gas grid or a gas grid in general.

For applications other than stationary, a separate storage system will be needed if SOFC systems are to be used in them. For example, a marine SOFC system using LNG as the fuel will need storage tanks suitable for LNG and so on.

It will be great, if all fuels mentioned in Fig. 5 can directly be fed into the fuel electrode of the SOFC system. Only then can the SOFC be truly called fuel flexible. However, this is not practically possible because many of these fuels have heavy hydrocarbons and other components which will either poison the catalyst, block the active fuel oxidation sites or simply degrade the fuel electrode. Hence, all fuels except hydrogen and methane will need a reformer or a partial reformer or a cracker prior to entering the fuel electrode of the SOFC. Even with methane, a proper amount of steam to carbon ratio should be maintained, typically between 2.5 and 5. What this means is a sufficient quantity of steam must be supplied along with methane to enable reforming of methane else the carbon species in methane will degrade the fuel electrode.

Aviation or aircrafts are currently not included in the above figure because electrification in this industry/application is still at a very nascent stage, and the use of fuel cells and batteries is either in a concept phase or prototype phase.

## 4 SOFC System Description

In this section, two typical SOFC systems, one operating with hydrogen as fuel, and the other operating with methane as fuel is described. This gives the reader an idea of how system architectures can be developed for the end application.

### *4.1 SOFC System Operating with Hydrogen as Fuel*

The schematic of an SOFC system operating with hydrogen as the fuel is shown in Fig. 6 and a corresponding heat exchanger network for internal heat recovery is shown in Fig. 7. Please note that this is one way to design the system and internal heat recovery. A system designer can integrate the components in another way too and that architecture might look different from that of Fig. 6.

The system architecture operating with hydrogen as fuel is quite simple. Besides the SOFC stack, the other BoP (Balance of Plant) components needed for the system are:

- Steam generator (optional)
- Fuel mixer
- Air compressor
- Fuel preheater

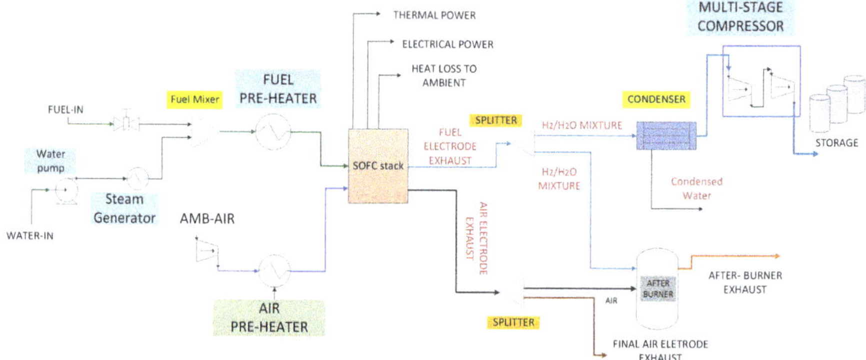


**Fig. 6** An SOFC system operating with hydrogen as fuel. (*AMB-Air* ambient air)

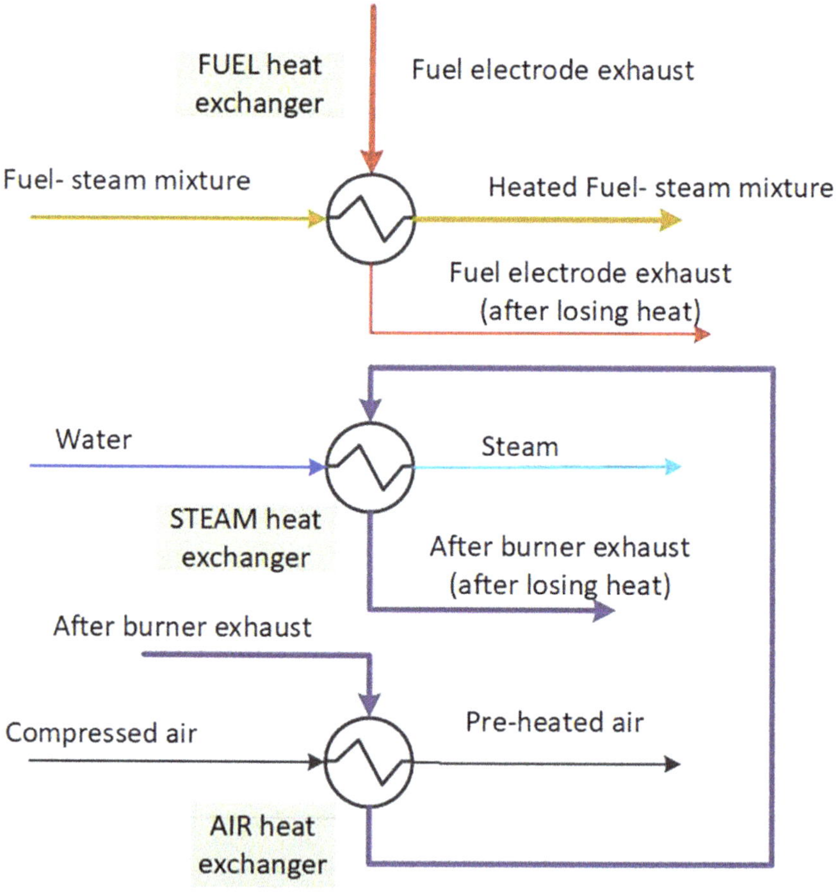

**Fig. 7** One possible heat exchanger network for internal heat integration

- Air preheater
- Fuel electrode exhaust splitter (optional)
- Air electrode exhaust splitter (optional)
- After burner (optional)
- Condenser
- Multi-stage compressor (optional)

Both the fuel and air need to be preheated prior to entering the stack. The usual rule of thumb is the streams need to be heated to a temperature of at least 50 K below the stack operating temperature. For example, if the stack is operating at 1073 K then the streams must be heated to at least 1023 K (or higher) prior to entering the stack. The fuel is electrochemically combusted in the stack and the exhaust from the air electrode and fuel electrode is either combusted in an afterburner and the resulting stream used for internal heat exchange prior to venting it out to the environment or the fuel electrode exhaust is recirculated back to the fuel electrode inlet and the air electrode exhaust just cooled and vented out.

Some of the BoP components in the above list is marked as optional and may not be needed. This totally depends on the application intended for which the system will be used for. The internal heat recovery may not be needed if the system is coupled to residual/waste heat processes which can potentially be coupled. Internal heat recovery plays an important role in standalone isolated systems.

## 4.2 SOFC System Operating with Methane as Fuel

The architecture of an SOFC system operating with methane as fuel is shown in Fig. 8. Methane might be a transition fuel until a complete hydrogen economy kicks in. The BoP components needed in addition to the ones mentioned for a hydrogen-based system are as follows:

- Fuel preheater prior to reformer
- External methane reformer

The system architecture is a little more complex when compared to a hydrogen-based system. This is because methane has to be reformed and converted to a fuel mixture which can be electrochemically combusted. The reforming process can either be external, as shown in Fig. 8 or can be internal, within the SOFC stack. The reforming reactions are given in Eqs. (3–5). As reforming is an endothermic process, heat needs to be supplied for the process to occur. In case of external reforming, heat has to be supplied to this reactor. This can be done by internal heat recovery from the SOFC system or from another process which rejects heat. With internal reforming, the heat produced by the stack can be used for the reforming reactions, but this requires a delicate thermal balance.

The remainder of the system operation is similar to the one described for the hydrogen-based one.

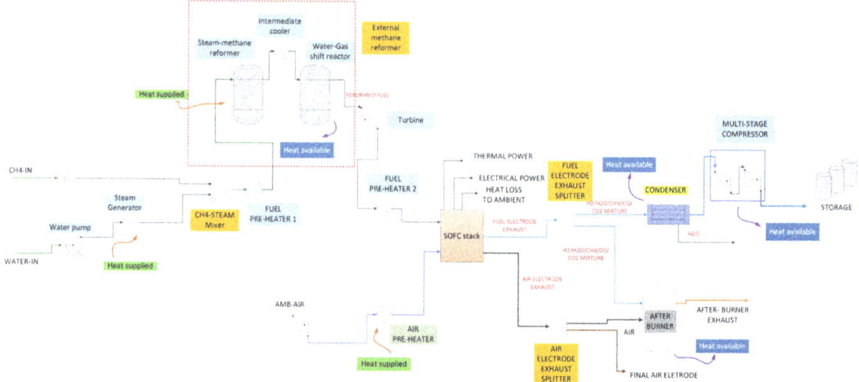


**Fig. 8** An SOFC system operating with methane as fuel, methane is externally reformed. (*AMB-air* ambient air)

## 5 Needs of the Aviation Industry

The growth in the aerospace industry has made globalization possible, making air travel affordable for the common public. However, this increased growth has also led to increased emissions from aircrafts and its related operations on the ground. The aviation sector contributes ~3.5% of all emissions related to climate change [5]. As a result of this, almost all airlines and aircraft manufacturers are looking for ways to cut down emissions and electrify systems (on both air and ground). Technologies developed in the aeronautical/aerospace industry are usually catalysts for innovation and spill-over into other economic and technological sectors. Thus, it is essential to develop alternate electrified solutions for powering aircrafts, even a tiny bit of electrification will help in reducing the emissions from the aircraft. Exploring alternate options for aircraft ground operations and also for different needs of airports will go a long way in reducing emissions from the aviation industry in general.

Note: The terms '*aircraft*', '*aviation*' and '*aerospace*' are used interchangeably in this section, all referring to air travel or related to air travel.

### 5.1 *Fuels Used on Aircrafts*

The commonly used fuels used in aviation are:

- Jet fuel (Jet A-1 or JP 1A, kerosene)
- Kerosene–gasoline mixture (Jet B)
- Aviation gasoline (avgas)
- Biokerosene

*Jet-fuel* is the most commonly used fuel in civil aviation. It is called JP 1A and is a type of kerosene which is obtained from careful refining of light petroleum. JP 1A has a flash point of 38 °C and a freezing point of −47 °C. *Jet B* also known as JP 4 is primarily used in military jets. It is a blend of 65% gasoline and 35% kerosene and has a flash point of 20 °C and a freezing point of −72 °C. *Aviation gasoline* is a kind of leaded fuel with a high octane number and used primarily on sports aircrafts and private aircrafts which have piston engines. As most aviation fuels are based on fossil fuels, there is increasing ongoing research to make them eco-friendly. *Biokerosene* can be made from algae and other biofuels can be made from jatropha or camelina oil. Biokerosene is basically a mixture of kerosene and biofuels and is still in the trial stage for use in different aircrafts. It is not completely fossil free and will still have some carbon footprint albeit lower. However, with developments and advancements in carbon capture technology, biofuels could be widely adopted. The reasons are one they can be used with conventional systems, requiring minimal modification, and the other is that the carbon emissions given out can be captured by the advanced carbon capture technologies. Hence, both performance and emission reduction (even net zero) are guaranteed.

Hydrogen as an aviation fuel is still a long way ahead. Energy density, volume, safety and other factors need to be proven. Liquified hydrogen has only 30% of the energy density of kerosene. Previous chapters in this book have extensively discussed hydrogen as a potential fuel. The reader is kindly requested to go back to those chapters for a detailed explanation.

## 5.2 *Major Components on Aircraft*

Fuel cells are increasingly competing with batteries and internal combustion engines to become the main propulsion source in almost all modes of transportation, and aviation or aircrafts are no exception. There are two options for using fuel cells on airplanes: (1) for propulsion and (2) as electrical energy generator. The former is needed during take-off, cruise and taxiing while the latter is needed to supply auxiliary loads on the aircraft.

The major components on a typical civilian aircraft are:

(a) Gas turbines—mainly used for propulsion
(b) RAT (Ram air turbine)—used for providing emergency power
(c) APU—this is fitted to the tail cone of the airplane
(d) Battery pack

Figure 9 shows the schematic of the location of these components on a civilian aircraft. Current APU on board aircrafts is a small gas turbine designed to supply part of the aircraft's electrical and pneumatic loads. Its primary purpose is to start the aircraft's main engines, and its secondary purpose is to provide electrical and pneumatic power during ground operations. It can also function as a turbine power

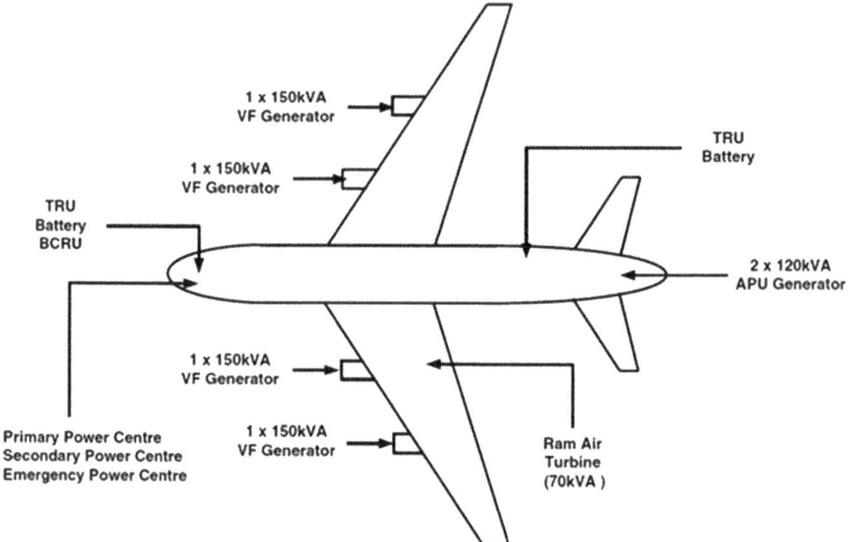


**Fig. 9** Schematic showing the different components on a civilian aircraft [6]

supply and emergency starter system. It is mainly used when the aircraft is grounded and during landing and take-off routines but is switched off at cruising altitudes.

The RAT is primarily used in case of an emergency where both the gas turbines and the APU have failed. The RAT is basically a small wind turbine which functions based on the air pressure, generating power to keep critical flight systems, controls and instrumentation in operation.

## *5.3 Potential Application of SOFCs in Aviation Industry*

The aviation industry (manufacturing, product development and system deployment) caters to three main centres of use viz. *airports*, *aircrafts* and *space applications*. All equipment, components and instrumentation developed and produced makes its way to one of these places. Among the three, space applications are quite niche and require extremely robust and effective solutions and systems.

In this section, first the current needs for power generation equipment and power sources at these centres of use will be discussed and then the potential use of SOFCs in these applications will be discussed.

## At Airports

The equipment needed and used at the airports for generating power is as follows:

- Ground power infrastructure at airports
- Small tractors
- Emergency power generators
- Push back tugs
- Large tractors
- Passenger transport bus (terminal to terminal)
- Passenger transport buggies (intra terminal)

Besides the above, power will also be needed at the airport terminal (passenger side) for lighting, heating and cooling (air conditioning) and operating other appliances such as baggage handling systems and terminal bridges. Power will also be needed for airport terminal airside, and these include runway lighting and auxiliary power units.

Power for airport terminals (passenger side) is mostly taken from the electricity grid. Hence, all stationary systems on the passenger side of the airport can be grid connected. If the grid electricity is decarbonized, then automatically all these systems are also decarbonized. It is the mobile systems on the passenger side of the airport which has a lot of scope for improvement as most of these systems currently run on internal combustion engines.

Among the bullet points mentioned above, the most promising areas where SOFCs can be deployed are *ground power infrastructure* (in case not connected to grid) *and emergency power generators*. For most of the other bullet points either a battery-based solution or a battery plus PEFC (Polymer Electrolyte Fuel Cell) solution would suffice. This will then completely decarbonize all operations on the passenger side of the airport. Passenger transport buggies are completely battery operated in almost all airports of the world.

On the air side of the airport terminal which can be a few miles away, either a grid connected solution or a grid independent solution will be needed. This depends on the airport and the way it was designed and built. In case the air side is not connected to the grid, then SOFC systems can be deployed to power runway lighting and also to provide any emergency or auxiliary power needed in case aircrafts get stranded on the runaway. This solution will be far cleaner and electrically efficient when compared to diesel-based generators.

## On Aircrafts

The major power generation components on the aircraft was mentioned in the previous section. Here, it will be imperative to first chalk out the power needs on the aircraft and then discuss potential applications of SOFCs as an alternative.

## Power Needs on Aircrafts

It is essential to get an overview on the power needs on board different aircrafts in order to design and develop electrification systems. Since the type and size of aircrafts are diverse and numerous, it will not be useful to lay down every single aircraft with its specific power requirements but instead group them in different categories and provide a power range for each of those categories.

Aircrafts in general can be categorized into *light*, *narrow body*, *wide body* and *UAV drones* as shown in Fig. 10. The power needed by an aircraft also depends on its flight stage—taxiing, ground operation, take off, cruise and descent. The power requirement range for the different types of aircrafts is summarized in Table 1.

A light aircraft has a maximum gross take-off weight of 5670 kg. Most production-certified light aircraft have engine powers in the range of 90–260 kW. Some models with a higher take-off weight have engine powers in the range of 300–900 kW. Narrow body aircrafts seats up to 295 passengers, some examples include the Airbus A320 and A321, Boeing B737-800 and B757-200. These aircrafts can do a maximum flying time of up to 9 h. Most engines used on the narrow body aircrafts provides a thrust between 67 and 160 kN, and typically two engines are used. This translates to a maximum power of 13 MW per engine. In some cases, this could also go up to 30 MW. Wide body aircrafts can seat up to 850 passengers. Boeing 747 and 777X, Airbus A380 falls under this category. In wide body aircrafts too, two engines are used, and these engines provide a thrust between 360 and 510 kN, which translates to 40 MW of maximum power, and this can go up to 100 MW in some cases. Any aircraft needs maximum power only during take-off, and hence, the engines have to be designed accordingly.

UAVs on the other hand need anywhere between 5 and 240 kW of power. Of course, they too have a cycle of operation with lift offs, travel and landing, but these power ranges can easily be met with an alternate source, in this instance by fuel cells. The advantage of having a fuel cell on the drone is that the flight duration can be extended in comparison to batteries, in some cases even up to 3 h [8, 9].

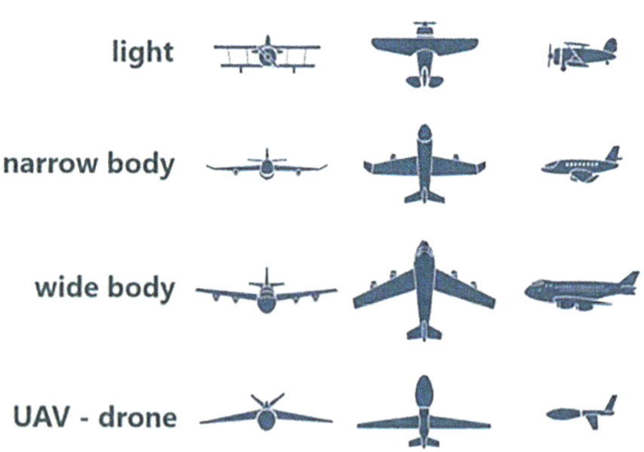


**Fig. 10** Classification of aircrafts into different types [7]

**Table 1** Power requirements for different types of aircrafts

| Type of aircraft | Power needed during take-off & climb | Power needed during cruise | Power needed during ground operations | Power needed during taxiing |
|---|---|---|---|---|
| Light | 120 kW | 55–75% of take-off | 2 kW | 10% of take-off. Depends on the surface beneath. Taxiing on grass needs more power than asphalt |
| Narrow body | ~30 MW | 55–90% of take-off | 40 kW electricity +90 kW air conditioning | ~7% of take-off (depends on size of aircraft and engine) |
| Wide body | ~80–100 MW | 75–90% of take off | 80 kW electricity +200 kW air conditioning | ~7% of take-off (depends on size of aircraft and engine) |
| UAV drone | Depends on size, 5–240 kW | Depends on functions (e.g. radar, measurements) | | |

Note: The above numbers are an indication only. The reader is advised to research further on a specific kind or type of aircraft for the exact power needs

The power rating of APUs (auxiliary power units) is between 40 and 200 kW for commercial aircrafts and between 25 and 300 kW for helicopters. These power ranges can easily be met with fuel cells (both SOFCs and PEFCs) and might be a very good starting point for implementation in the aircraft electrical architecture. In automotive parlance, systems between 50 and 160 kW are classified as light duty and systems between 160 and 400 kW are classified as heavy duty. Both light duty and heavy duty SOFC systems are ready from a technological perspective, but there are certain engineering challenges that need to be solved in order to make them compatible for use in aeroplanes/aircrafts. For example, a 50 kW SOFC system would need roughly a space of $2.5 \times 2.8 \times 1.8\ m^3$. This is because some of the BoP (Balance of Plant) components of an SOFC system are large. This is very large and not acceptable for an aircraft application. However, this can be solved with advancements in engineering where compact high-performance components can be designed and developed. For example, further advances in micro-heat exchanger technology will drastically reduce the size of fuel and air pre-heaters within the SOFC system. Also, all fuel cell systems so far have only been certified for ground level use. If these systems are to be used for aircrafts, then they must be certified for air worthiness.

To give the reader further insights into the use of SOFC technology/systems on aircrafts, some literature from research studies are presented here. The literature is only presented as a flavour to spark the reader's interest and the reader is advised to access these papers individually for a more thorough understanding. Fernandes et al. [10] have written a review paper on '*SOFC APU systems for aircrafts*', where they discuss in depth the different works published especially on this topic. The focus of their review paper is how SOFC APU systems could potentially be a replacement for conventional APU systems. In a more detailed industry linked study, researchers from Colorado School of Mines and United Technologies Research Centre looked into [11] how different architectures and system concepts can be designed with SOFC APUs. They concluded that significant fuel burn savings can be made during ground operations in comparison to in-flight mission segments. This points to the potential role that SOFCs can play when aircrafts touch the ground and power needs on the ground (during taxiing, parking, etc.) are to be met. Guillerm [12] presented a modelling study on how SOFC systems can be integrated in the aircraft electrical architecture. He found an optimum operating pressure of 3.3 bar linked to the system which gives optimal performance. Although this is just a modelling study, some insights on how much fuel can be saved, the system performance during the entire flight duration and the positioning of the system on the aircraft has been given. This provides further incentive for relevant stakeholders to look deeper into how these systems can be an alternate to conventional ones. Ji et al. [13] talk about an SOFC hybrid jet engine which operates under different modes. Mathematical modelling of the entire system is carried out under the different operating modes. The different operating modes are distinguished based on the fuel and air flows that are regulated to different parts of the system (e.g. afterburner, reformer, SOFC). Their objective was to study how the system would behave under these different conditions. Their study is more of a thermodynamic study where operating and design maps are drawn. This provides crucial data for people looking into building such systems.

What all these studies indicate is that it is indeed feasible to have an SOFC system replace the current conventional system, and it is just a matter of time when advances in SOFC technology will make it happen. The key criteria to be met to make that happen is laid out in these studies which can serve as inputs for the advancement of SOFC technology.

Besides power needs on board, all aircrafts are connected to ground power when stationed at the airport in order to avoid burning aviation fuel and also to prevent damage to surrounding infrastructure resulting from the blast of jet engines. Most of the major airports provide electrified power for aircrafts via the ground bridge also known as the passenger boarding bridge. However, at many of the airports, the ground power is still based on diesel gensets. Figure 11 shows the way ground power is supplied to aircrafts. The picture on the left shows the use of diesel gensets and the right shows how ground power supply is integrated into the boarding bridge.

To conclude, out of the four power generation components on the aircraft, it is envisaged that there is scope for use of SOFC based systems as an APU in the short term and as a hybrid propulsion unit in conjunction with gas turbines in the long term. The other two power sources viz. RAT and battery will remain as such with little or no modification at all as there is not much of a technical case or need to replace these.

**Fig. 11** Power being supplied to aircrafts when grounded or stationed at airports [14, 15]

## Space Applications

Systems and components used for space applications have a very specific use case and extremely niche. For space applications, there is always a need for (1) electrical power, (2) oxygen and (3) water on board the space vehicle. The electrical power needs can easily be met with batteries along with solar panels. Solar energy is available in copious amounts and with the right battery sizing, electrical power can be made available all the time. The solar panels will constantly charge the battery, and thus, there is very little risk of power unavailability. Fuel cells can also be used for electrical power, but they are limited by the fuel storage on board the spaceship or the space vehicle unless this fuel can be produced in some way on board. One concept that has been propagating is to use a fuel cell-electrolyser combination. The way this system is envisaged to work is as follows: the fuel cell system (with its initial storage tanks) produces the required power and a certain quantity of water. This produced water is captured and stored. The water is then electrolysed by the electrolyser system to produce the fuel, which is to be used during fuel cell operation mode. The energy needed for water electrolysis is provided by the solar panels.

Taking into account that room for a system is limited on board the spaceship or the space vehicle, the fuel cell system must be as compact as possible. The fuel cell–electrolyser combination concept described above is limited by (1) the fuels which can be used directly within the fuel cell and (2) the whole cyclic or circular operation. Only two fuels can be directly used within a fuel cell as mentioned in section 3 of this chapter viz. hydrogen and methane. For all other fuels, an external reformer and gas cleaning device will be needed. Out of the two fuels, methane cannot be directly used with a PEFC system but can be used in an SOFC system. Hence, if a PEM-based fuel cell–electrolyser combination is used, then hydrogen is the fuel of choice. Even though methane can be used directly in an SOFC system, it is limited by the second point viz. the circular approach. The circular approach cannot be guaranteed if methane is used as the fuel unless there is an external methanation reactor included in the SOC system. Including an external methanation reactor will add complexity to the system and eat up more space. However, there is a way for the circular approach to work for SOFC-SOEC (Solid Oxide Fuel Cell–Solid Oxide Electrolyser Cell)-based system. The storage tanks will initially be methane to start with but will gradually be replaced with hydrogen. This is because the SOEC will carry out steam electrolysis and hydrogen will be produced which can then be used in the fuel cell operation mode. The challenge though will be the generation of steam from the water collected. Now considering that the SOFC system will eventually work with hydrogen after a certain period of time, it does not make sense to start off with methane storage tanks. Hence, hydrogen will be the fuel of choice for both technology types—PEFC or SOC (Solid Oxide Cell).

The advantage of fuel cells is the generation of water which can be used for various purposes inside the spacecraft. If batteries are used, then water must be carried for the entire mission duration and should be used with caution, resulting in a large storage tank. With fuel cells, water can be produced anytime when needed, and the trade-off will be between the fuel consumed and storage tank needed for water. A smaller storage tank might be sufficient when fuel cells are used. With the right optimisation and sizing, the cyclic approach can be achieved where fuel is generated from water, and water is generated as a by-product from fuel cell mode. Part of this water can be used for other purposes on board the aircraft and part of which can be used for electrolysis.

SOCs in electrolyser mode can also be used for $CO_2$ electrolysis, and the produced fuel can be stored and later used during fuel cell operation mode. They are the only electrolysis technology that has reached a TRL (Technology Readiness Level) of 8 for $CO_2$ electrolysis. The basic reactions occurring at the electrodes are as follows:

$$CO_2 + 2e^- \rightarrow CO + O^{2-} \tag{6}$$

$$O^{2-} \rightarrow \frac{1}{2}O_2 + 2e^- \tag{7}$$

The produced CO is a potential fuel for SOFC operation. The other by-product which is pure oxygen is quite valuable on board the spacecraft and can be used by astronauts or to supply the air electrode of the SOFC or for any other purpose. This kind of a system will find application on an environment where $CO_2$ is abundant such as the Martian atmosphere.

Thus, to conclude, systems based on Solid Oxide Cells are a very good candidate for the generation of electrical power, water and oxygen on board a spaceship or a spacecraft.

## 6 Concluding Remarks and Perspectives

In the previous sections of this chapter, the reader was introduced to the basics of solid oxide fuel cells, the different cell designs and cell architectures possible, the fuels that can be used with SOFCs, two different system architectures—one operating with hydrogen as the fuel and the other operating with methane as the fuel and the potential applications of SOFC-based systems in the aerospace industry.

Current state of the art SOFC stacks have a specific power between 0.26 and 0.68 kW/kg [16]. This specific power goes down by ~80% when considered on a system level. This is because of the additional weight when considering all the BoP components. Thus, on a system level, one can expect a specific power between 0.05 and 0.14 kW/kg. Now this specific power is way lesser than gas turbines used on aircrafts. The goal is to get to a specific power of 1 kW/kg for SOFC stacks and to a specific power of 0.5 kW/kg for SOFC systems [17]. NASA's Glenn research centre [18] has developed an SOFC stack with a specific power of 2.5 kW/kg. This stack design is already about five times better than the current state of the art stacks. With these kind of quantum leaps, it looks even more promising that SOFCs will play a greater role in the aerospace industry.

As mentioned above, since the state of the art SOFC stacks have a low specific power, adding an SOFC system to the aircraft would increase the system weight, but many studies have also shown that system performance is increased and emissions are reduced [19, 20]. Thus, the trade-off comes at a design stage where the aircraft manufacturer has to make a choice between fuel consumption, weight, system complexity and performance. With the current state of the art technology, SOFCs may not be the best choice for aircraft propulsion systems; however, there is ample scope for them to be used as electrical generators.

The environmental conditions around which aircraft systems have to work are quite harsh. Cruise conditions are at an altitude of 36,000 ft. (~11 km). Here the ambient pressure is around 0.2 bar, way below the atmospheric and the ambient temperature is around −55 °C. Spacecrafts or space vehicles will encounter even harsher conditions—including zero gravity conditions and lack of air. Thus, if fuel cell systems are to be deployed, then increased airflow is required to obtain the same amount of oxygen which is needed at sea level. Also, rigorous testing of fuel cell systems must be carried out where actual flight conditions or space atmosphere is emulated, and this will warrant specialized test beds. One to one replacement of aircraft powertrains may not be possible or even feasible and might warrant completely new aircraft architectures.

## 6.1 Why SOFCs Will Be a Good Fit?

SOFC technology is envisaged to play an important role in air transportation in the new millennium. In any transport application, the choice of fuel will ultimately decide the technology that gets engineered for it. In the aerospace industry/sector, there is a requirement for alternate solutions, for both propulsion and energy storage, which are (1) high in energy efficiency, (2) low on emissions and (3) have excellent reliability. So far, neither fuel cells nor batteries have been able to match the performance of conventional systems, but their performance is getting closer to the conventional systems every year with advances in technology.

Even though hydrogen is touted to be the fuel for the future, for aircrafts and other aerospace related applications, there is a chance it may not become the fuel of choice for aviation, and in such a scenario, PEFC technology may not get adopted for the aviation industry. This is because hydrogen generation is still not cost-effective and the price per kilogram of hydrogen is still not competitive with conventional fuels. Possibilities of complete battery-driven solutions also look bleak just because of the sheer volume of batteries that will be required to reach specific power levels of conventional gas turbines. In meanwhile, processes to generate conventional fuels are also getting cleaner, with advances in carbon capture technology and other related things. However, one can say that battery-based solutions have been proven on very small aircrafts (unmanned, one and two seater planes).

Considering the above, SOFCs seem to be a good candidate for a number of reasons, some of which have been mentioned in chapter "Hydrogen Storage Technology for Aerial Vehicles", fuel flexibility being one of the key advantages. For the same stack power, an SOFC system would weigh less when compared to a PEFC system. This is true only for fuels other than hydrogen because there is a need for reforming and gas processing equipment and that makes the PEFC system bulkier. Also, SOFC systems can be used in conjunction with current gas turbine systems and with advances in technology, it is very much likely that the future aircraft engines will be some kind of a hybrid (gas turbine–SOFC combination).

Current APUs based on gas turbines have a best-case efficiency of 40% during flight and an efficiency of 20% on land. These numbers can be significantly boosted by 20% and 40%, respectively, if an SOFC stack is used on board (an SOFC stack has an electrical efficiency of about 60%).

## 6.2 Perspectives

Safety is of utmost importance when developing solutions for aircrafts. There cannot be room for any kind of error and provisions should be in place for a second or even third level of systems if the primary functioning system fails. The conventional systems in the aerospace sector already have these in place, the same should also be implemented when replacing conventional systems with alternate ones.

The aerospace industry helps sustain about $2.4 trillion in economic activity, and the figure is expected to increase to $6 trillion by 2040 [21]. Hence, electrification in this industry presents a huge business potential-provided appropriate solutions are available and engineered. Electromobility currently exists for single and twin seater categories and operation at higher voltage results in reduced mass of wiring. This must be taken into account when going for electrified solutions on board the aircraft. The trade-offs for electrification with current state of the art technologies include reduction in payload capacity. Now whether this is acceptable or not is the critical question which has to be decided by the experts in the industry.

What is the aviation fuel of the future? The answer to this question is not readily available. Each fuel will require different components integrated in a unique architecture—each having its own benefits and challenges. Hydrocarbon-based fuels (fossil-based or bio-based) are going to linger around in the aviation sector for at least a couple of decades, but at the same time, emissions have to be brought down drastically. The only logical step in such a case would be to look for alternate high energy efficient solutions that can use the fuels that are available as of today. Technologies must not only be developed for the long-term future (three or more decades) but also for the short-term future (one or two decades).

A tentative roadmap up to and just beyond 2050 drawn for the aerospace industry is outlined below [21]:

- Aircraft movements will become emission free when taxiing by 2050. This means as soon as the plane touches the ground and until it parks itself at the terminal, it should run on electrified power in order to have zero emissions. This can be accomplished via either batteries or fuel cells. For light aircrafts, the power requirements during taxiing can still be accomplished via electrification technologies but for narrow and wide body aircrafts, the power required during taxiing is still in the MW range which may be hard to accomplish even by 2050.
- Commercial aviation (50–70 seats) and equivalent cargo < 7 tons:
  - By 2025, it might become mandatory to electrify APUs and all non-propulsive components. Emergence of de-centralized architecture is to be expected unlike the current systems. By 2035, it is expected to have 30% hybridization for power and 10% hybridization for energy, with implementation of distribution propulsion. By 2055, it is expected to have full electric aircrafts for short haul flights—for both propulsion and power. Batteries and fuel cells are expected to play a major role in electrification with 50% hybridization of power and 20% for energy.
- Commercial aviation (>70 seats) and equivalent cargo > 7 tons:
  - As systems get bigger, so do engineering challenges. In this class of aircrafts, it is only by 2035 that electrification of APU and all non-propulsive systems is expected. By 2055, electrification of propulsion is also expected with 30% hybridization for power and 10% hybridization for stored energy.
- Personal aviation and drones:

  - These aircrafts tend to be smaller. Hence, by 2025, more than 50% hybridization is expected in the propulsion. The flight range is expected to be short, less than 200 km. By 2035, full electric aircrafts which can do 200 km are expected to fly around and by 2055 full electric aircrafts which can do 500 km are expected.

- Airports (passenger side and air side):
  - By 2025, all ground-based services and support vehicles are to be electrified, and airports must have the infrastructure for providing charging facilities. By 2035, it is envisaged that initial concepts for induction charging will be introduced, and airports will have facilities to charge small airplanes (50–70 seats). By 2055, complete electrification of every aspect of the airport is to be expected and facilities to charge any kind of planes should be available.

If alternate solutions are to be implemented on board aircrafts, an energy storage system is needed. An energy density of 750 Wh/kg will provide sufficient competency. Figure 12 shows the spot on the graph of MTBF (mean time between failures) vs. load cycle where aircrafts with fuel cell technology would fit.

As one can see the load cycle is less stringent when compared to the automotive sector and thus using SOFCs over PEFCs can be more suitable when it comes down to such applications. Another advantage is that fuel cells in general provide direct current which reduces the number of inverters and converters needed for power conversion. This is an indirect benefit which will help reduce weight of miscellaneous electrical/ electronic accessories.

With increasing electrification on board planes, there will also arise the need for infrastructure for charging electric planes at airports. Stationary power generation using SOFCs could be one solution in places where airports are unable to provide the required infrastructure. This could be another market for stationary SOFC

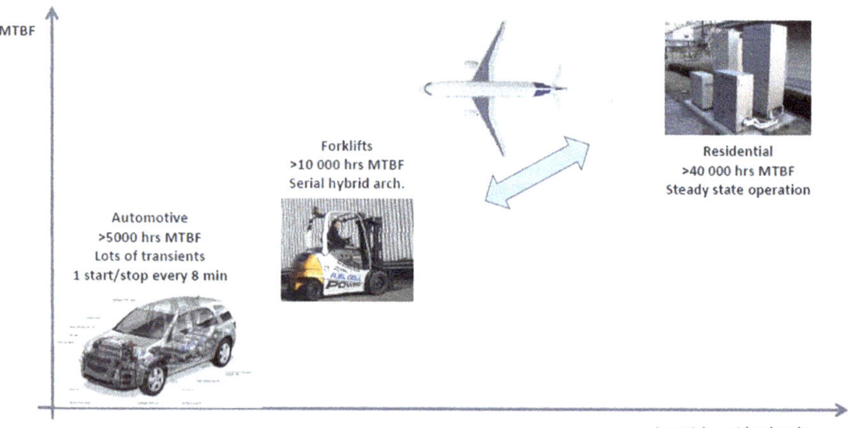


**Fig. 12** Schematic showing where airplanes fit in if fuel cells are used [22]

systems. Such systems will help minimize fuel consumption, increase energy efficiency and reduce noise and $CO_2$ emissions at airports.

Some of the key questions that should be dwelled into if fuel cells are to make inroads in the aerospace sector are:

1. When taxiing, emission should be zero. How can fuel cells help here? What is the taxiing operation cycle like?
2. Small aircrafts which can travel up to 200 km might require 1 MW per engine. Can this be electrified? What are the most promising hybridisation architectures?
3. How can the learnings from small aircrafts be extended to large aircrafts?
4. How can fuel cell systems be engineered for APUs in order to make them zero emission? Engineering and manufacturing options need to be explored.

## References


1. *Technology Roadmap, Hydrogen and Fuel Cells*, International Energy Agency, 2015.
2. *Hydrogen and Fuel Cells: Opportunities for Growth – A roadmap for the UK*; David Hart, Jo Howes, Ben Madden, Edward Boyd, November 2016.
3. *Development of highly efficient planar solid oxide fuel cells*; Kazuhiko Nozawa, Himeko Orui, Takeshi Komatsu, Reiichi Chiba and Hajime Arai, Special feature: NTT technologies for the environment.
4. *Solid Oxide Fuel Cell (SOFC) technology for greener airplanes*; Larry Chick, Mike Rinker; Energy Materials Group, Pacific Northwest National Laboratory, September 2010.
5. https://en.wikipedia.org/wiki/Environmental_impact_of_aviation
6. Book: Civil Avionics Systems; ISBN 10:1118341805
7. https://www.istockphoto.com/de/vektor/verschiedene-arten-von-flugzeugen-gm538891218-95952109
8. https://www.dti.dk/fuel-cells-extend-flight-time-on-drones/37268
9. https://blog.ballard.com/commercial-uav-flight-times
10. '*SOFC-APU systems for aircraft: A Review*'; Int. Journal of Hydrogen Energy; vol.43; 16311–16333 (2018)
11. '*System architectures for Solid Oxide Fuel Cell-based Auxiliary Power Units in future commercial aircraft Applications*'; Journal of Fuel Cell Science and Technology; vol 6; 031015-2 (2009)
12. '*Assessment of a Solid Oxide Fuel Cell powering a full electric aircraft subsystem architecture*'; Department of Aerospace and Vehicle Engineering, Royal Institute of Technology, Stockholm, Sweden
13. '*Performance characteristics of a solid oxide fuel cell hybrid jet engine under different operating modes*'; Aerospace Science and Technology; vol 105; 106027 (2020)
14. https://itwgse.com/
15. https://www.jbtc.com
16. '*Design and performance of compact Air breathing jet hybrid-electric engine coupled with Solid Oxide Fuel Cells*'; doi: 10.3389/fenrg.2020.613205; Frontiers in Energy Research; Vol 8, Article 613205
17. '*SOFC development for aircraft application*' – Presentation by G Schiller; 1st International Workshop on SOFCs: "How to bridge the gap from R&D to market"?; Quebec, 15 May 2005
18. '*High Power Density Solid Oxide Fuel Cell*' – NASA Glenn Research Centre

19. 'Investigation of two hybrid aircraft propulsion and powering systems using alternative fuels'; Energy vol. 232; 121037 (2021)
20. '*All electric commercial aviation with solid oxide fuel cell gas turbine battery hybrids*'; Applied Energy vol. 265; 114787 (2020)
21. *Electrification of the transport system – Studies and reports*; Directorate General for Research and Innovation; Smart, Green and Integrated Transport, European Union 2017.
22. *The Airbus Fuel Cell approach*; EYVE Airbus Fuel Cell Systems Engineering; presentation by Barnaby Law – Head of Department 'Integrated Fuel Cell Technology'. September 2012

# Index




C. O. Colpan, A. Kovač (eds.), *Fuel Cell and Hydrogen Technologies in Aviation*, Sustainable Aviation, https://doi.org/10.1007/978-3-030-99018-3

## L



## M



## N



## O



## P



## R

## S



## T



## U



## V



## W



## Y



## Z

GPSR Compliance
The European Union's (EU) General Product Safety Regulation (GPSR) is a set of rules that requires consumer products to be safe and our obligations to ensure this.

If you have any concerns about our products, you can contact us on

ProductSafety@springernature.com

In case Publisher is established outside the EU, the EU authorized representative is:

Springer Nature Customer Service Center GmbH
Europaplatz 3
69115 Heidelberg, Germany

www.ingramcontent.com/pod-product-compliance
Ingram Content Group UK Ltd.
Pitfield, Milton Keynes, MK11 3LW, UK
UKHW021834270726
14058UKWH00001B/141
* 9 7 8 3 0 3 0 9 9 0 2 0 6 *